D1583026

EARTH, MOON, AND PLANETS

THE HARVARD BOOKS ON ASTRONOMY

Edited by HARLOW SHAPLEY *and* CECILIA PAYNE-GAPOSCHKIN

ATOMS, STARS, AND NEBULAE
L. H. Aller

OUR SUN
Donald Menzel

GALAXIES
Harlow Shapley

BETWEEN THE PLANETS
Fletcher G. Watson

STARS IN THE MAKING
Cecilia Payne-Gaposchkin

THE MILKY WAY
Bart J. Bok and Priscilla F. Bok

TOOLS OF THE ASTRONOMER
G. R. Miczaika and William M. Sinton

The Earth just after third quarter as seen from the Moon. (Courtesy U.S. National Aeronautics and Space Administration.)

Fred L. Whipple

EARTH, MOON, AND PLANETS

THIRD EDITION

HARVARD UNIVERSITY PRESS

Cambridge, Massachusetts

Third Printing, Third Edition, 1970

Distributed in Great Britain by Oxford University Press, London

Library of Congress Catalog Card Number 68-21987

SBN 674-22400-0

Printed in the United States of America

This book was photocomposed by Graphic Services, Inc., York, Pennsylvania; printed and bound by Colonial Press, Clinton, Massachusetts.

Preface

In revising this book I am struck, not by the increase of knowledge concerning the planets that we have gained in two decades, but rather by the greater amount that we may expect to gain in a far shorter time. The advent of radio astronomy, satellites, interplanetary probes, and orbiting telescopes gives promise of the attainment of tremendous factual knowledge about the solar system, leading to a depth of understanding we could not hope to gain from beneath the Earth's cloudy, tremulous, and largely opaque atmosphere. The many problems of the solar system and its evolution take on renewed interest and astronomical vitality as one senses their impending solution.

In revising the text I have attempted to keep the spirit of the original book and not to be overwhelmed by the amazing technical progress that is in process of widening our horizons so markedly. The technical level of the book remains unchanged, although young readers will find its level relatively easier than it was to students of their age two decades ago. Fortunately, the technical and scientific sophistication of our high school and college students has risen enormously and is still rising. Even so, all mathematics is still avoided in this book; the scientific methods and processes are described in the simplest possible terms for those without specialists' training in science.

I am indebted to I. Bernard Cohen for wise guidance in historical matters and to Gerard P. Kuiper for special photographs of the Moon. I am especially grateful to E. C. Slipher for the use of photographs from the Lowell Observatory, many of which stand today, as they did two decades ago, at the presently attainable zenith of planetary photography.

February 1963   FRED L. WHIPPLE

Note to the Third Edition

As predicted in the preface above, the increase in our knowledge of the solar system during the five years since my revision of this book in 1963 exceeds that in the previous two decades. The program of the U.S. National Aeronautics and Space Administration, including certain of their centers and the Jet Propulsion Laboratory of the California Institute of Technology, and the space program of the U.S.S.R. have contributed a lion's share to this rapid progress. The impetus of these space programs has also carried over to the ground-based studies of the Moon and planets, and these studies now move at an accelerated pace. The advances in radio and radar astronomy are striking. May the acceleration continue!

January 1968   FRED L. WHIPPLE

Contents

PHOTOGRAPHS

Please note that in this book the photographic illustrations of celestial objects are intentionally inverted. The south direction is at the top and the north direction at the bottom. Since the astronomical telescope ordinarily inverts the image, the celestial objects are portrayed here with the same orientation that they present telescopically.

Earth, Moon, and Planets

Introducing the Planets

The five bright planets have been known to man for many thousands of years, but in antiquity they were regarded as mysterious celestial deities whose very motions seemed to reflect the caprices of super-human beings. The old Greek and Roman legends are well known. Mars was the god of war, Venus the goddess of love, while Mercury was a sort of messenger boy. Today, the situation has changed. Man-made vehicles now circle the planets and even land on them. These massive planetary spheres of iron, stone, and gas have real significance in our lives and in our thinking. We have the opportunity to search for living organisms that may have developed independently of the earth. The mathematical calculation of orbital paths for our space vehicles is a practical engineering problem. The composition of the atmospheres, the temperatures, and the nature of the planets occupy the attention of many people, not only scientists but engineers, explorers, astronauts, politicians, military men, businessmen, and others. Thus, astronomy has moved from its ivory tower into the market place and the planets have become our next-door neighbors.

Each planet acquires more individuality and becomes more interesting as additional facts accumulate. Every year some of the problems of the preceding year are solved and new ones, once beyond the expectation of solution, appear within reach. To appreciate the current discoveries and deductions in planetary astronomy, we must be familiar with the fund of knowledge already accumulated.

To proceed toward an understanding of the solar family of planets, their dependents, and the other inhabitants of this realm, an introduction is first in order. As everyone knows, the process of meeting a numerous family *en masse* may be both exhilarating and confusing. We shall proceed quickly with the introductions and then spend some time with each member of the family to develop more intimate acquaintance.

The planets are really so small compared to the vast distances between them, and their reflected sunlight is so weak in comparison with the great brilliancy of the Sun, that all of them can never show to good advantage from any one location. As a vantage spot for observations, our present position on the Earth is actually quite satisfactory, except for the thick atmosphere above us. Since we must surmount this obstacle, we might as well, in imagination if not yet by space ship, go out farther from the Sun to about the distance of Jupiter. From there the inner part of the solar system is easily visible (Fig. 1). The orbits of the planets as sketched in the figure show first that the Sun is almost exactly in the center. The reason is very simple; the Sun possesses 99.866 percent of the entire mass of the system, so that by gravitational attraction it completely dominates the motions of the planets.

We notice next that the orbits lie almost in a plane, very close to

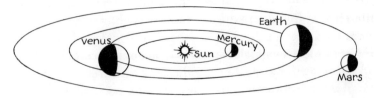

Fig. 1. Orbits of the inner planets about the Sun—a projection. The relative sizes of the planets are indicated, but on the scale used, the Sun's diameter would be 3 feet.

the *ecliptic,* the plane of the Earth's orbit about the Sun. This favoritism on the part of the planets in adopting a common plane of motion is probably not due to chance. Although no rigorous proof has been given, it is possible that Jupiter is responsible, because this planet is 318 times as massive as the Earth and possesses 0.7 of the combined mass of all the planets. Jupiter is certainly the master planet and by gravitational attraction may have regulated the orbits of the others. There is the more likely possibility, of course, that the planets were all formed in a plane—but we must investigate this matter later on.

Mercury, the smallest planet, moves in the smallest orbit of all, but one that is tipped from the common plane with an inclination of 7°, while the other inner planets keep within about 3° of the plane.

For measuring distances in the solar system we must use a larger unit than for distances on the Earth. The most convenient is the *astronomical unit* (A.U.), which equals the mean of the greatest and least distances from the Earth to the Sun, technically called the Earth's *mean distance.* This basic yardstick is some 92,956,000 miles long, according to astronomical and Venus radar measures. The distance is still uncertain by a few hundred miles.

The distance to the Sun is enormous in terms of ordinary distances on the Earth. An airplane moving with the velocity of sound, 750 miles per hour, would require 14 years for the trip (necessarily one-way), while a rocket moving at 5 miles per second would arrive in 7 months. If such distances seem large, remember that most of us spend our lives confined to one of the smaller planets of the solar system, and are denied the privilege of a "fuller" existence in the universe at large. The astronomical unit is actually much too small for conveniently listing the distances between the stars; a much larger unit, the distance light travels in a year, is often used for that purpose. This unit, as well as numerical constants for the planets, is given in Appendix 3. A convenient scheme for remembering the distances of the planets, Bode's law, is given in Appendix 1.

Mercury has a mean distance from the Sun of only 0.39 astronomical unit, Venus 0.72, the Earth 1.00, Mars 1.52, and Jupiter 5.20, a rather uniform sequence of increasing distances except for the large gap between Mars and Jupiter. In this gap we find thousands of small planets called *asteroids* that fill the space where a planet might well move (see Fig. 2 and note Fig. 7, later). These asteroids range from mountain size, a mile or so in diameter, up to Ceres which is about 480 miles across—comparable to a large island. Pallas comes second with a diameter of 304 miles and Vesta third, 240 miles. There are certainly no large asteroids that have not been discovered but there are many smaller ones—more than 50,000—that could be photographed with the larger telescopes. These fly-weight planets, although contributing a negligible part to the mass of the system (perhaps 1/500 of the Earth's mass), provide astronomers with

Fig. 2. The famous asteroid Eros leaves a trail on this time exposure as it moves through a field of stars. (Photograph by the Yerkes Observatory.)

a great amount of work in observation and calculation. They are fine test specimens for theories of various kinds and are assisting materially in finding the key to the origin of the solar system. Probably they were formed much in their present positions although a certain amount of collisional breakup has doubtlessly scattered them somewhat.

The planets themselves have much of the character of the ancient gods for whom they were named. Mercury is indeed swift and small, characteristic of a messenger. It requires only 88 days for a complete revolution about the Sun, less than one-fourth the length of our year. Its diameter is only 0.4 that of the Earth. Even this small diameter, 3025 miles, is enough greater than the diameter of Ceres to establish Mercury definitely as a planet rather than as a large asteroid. Radar shows that it rotates in about 59 days, probably exactly two-thirds of its period of revolution about the Sun. The planet is, unfortunately, so small and always remains so close to the Sun, as observed from the Earth, that surface markings are difficult to discern.

Venus is certainly the "sister" planet of the Earth. The diameter is almost identical (95 percent), the period of revolution about the Sun somewhat shorter (225 days), and the mass about 0.8 that of the Earth. Venus too is cloaked with a large atmosphere; it is this opaque atmosphere that hides the surface features so completely that we cannot see even the direction of rotation. From radar observations we judge the period to be long (Chapter 12), some 244 days, and *retrograde,* opposite to the direction of revolution. Radio data show that Venus is very hot, 600°F or more on its invisible surface. The actual observations of both of these inner planets are made difficult because we can see only part of the sunlit face at any time. When nearest to the Earth, Venus shows a thin crescent, as is the case for the Moon when new or old (Fig. 3), because the bodies are almost on a line between the Earth and the Sun. Figure 4 shows the various positions of Venus when the photographs of Fig. 5 were made. The various planetary configurations are named and defined in Appendix 2.

Mars may best be described as a pygmy Earth (one-half its diameter) with a thin atmosphere, distinct surface features, but no oceans. It moves more slowly about the Sun, in 687 days. Mars, however, boasts two moons or satellites, while Mercury and Venus have none. These two satellites are little more than medals for Mars, the god of war, since the larger, Phobos, is possibly

Fig. 3. The old Moon. (Photograph by the Lick Observatory.)

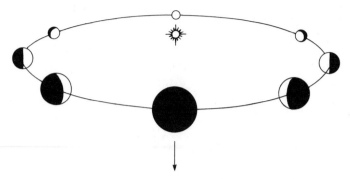

Fig. 4. The phases of Venus relative to the Earth. See Fig. 5 for corresponding photographs.

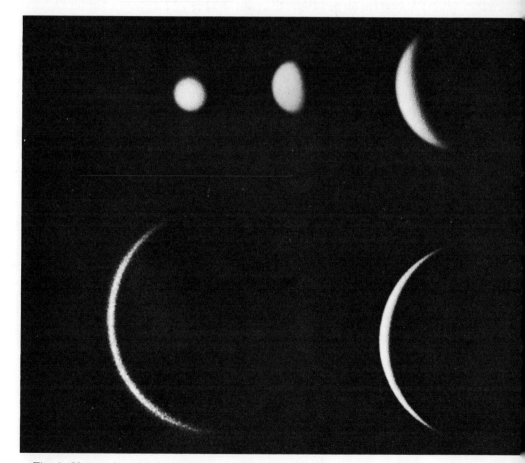

Fig. 5. Venus photographed with a constant magnification at various phases. (Photographs by E. C. Slipher, Lowell Observatory.)

only 5 miles across. Deimos, the fainter and therefore probably the smaller, may be only 3 miles in diameter. The existence of these Lilliputian satellites, strangely enough, was narrated by Jonathan Swift (1667–1745) in his *Gulliver's Travels* (Laputa, chap. 3) some 150 years before their discovery in 1877. According to Gulliver's account, the astronomers of the cloud island, Laputa, possessed small but most excellent telescopes and had "discovered two lesser stars, or satellites, which revolve about Mars; whereof the innermost is distant from the center of the primary planet exactly three of his diameters, and the outermost, five; the former revolves in the space of ten hours, and the latter in twenty-one and a half."

These periods of revolution are remarkably close to the truth, for Phobos revolves about Mars in 7 hours 39 minutes, while Deimos requires 30 hours 18 minutes. The mythical distances from the center of Mars are, however, too great; Phobos is distant only 1.4 diameters of the planet, and Deimos 3.5 diameters (Fig. 6). It would be enlightening to learn more of the Laputian discoveries, but Gulliver mentions only that the Laputians had "observed ninety-three different comets, and settled their periods with great exactness."

The rapid motion of Phobos makes this satellite unique in the solar system. Its period of revolution is less than the Martian day, 24 hours 37 minutes. As seen from the surface of Mars, Phobos would rise in the west and set in the east!

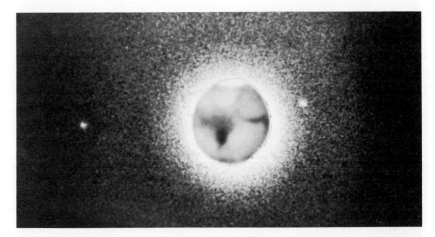

Fig. 6. Mars with its satellites Deimos (*left*) and Phobos (*right*). The image of the planet was overexposed in photographing the faint satellites, and was replaced by a normally exposed image. (Photograph by E. C. Slipher, Lowell Observatory.)

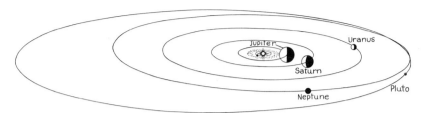

Fig. 7. Orbits of the outer planets about the Sun—a projection. Pluto passes within Neptune's orbit, but does not intersect it because of the inclination. Note the asteroids and the relatively small size of the orbit of Mars.

Before going on to the outermost planets, we note that the four planets Mercury, Venus, Earth, and Mars are really very much alike, of somewhat the same size, and all fairly dense, as though they were made of stone or iron. They are justly classed as the *terrestrial* planets because of their similarity to the Earth. Probably Pluto is much like a large well-frozen asteroid. Jupiter, Saturn, Uranus, and Neptune, on the other hand, are of an entirely different species, giants compared to the Earth, and only about as dense as water. Their orbits are shown in Fig. 7 as seen from beyond the distance of Pluto. On this small-scale chart the orbits of Fig. 1 are all crowded into a small region about the Sun.

Jupiter is conspicuous as the greatest of the planets. It has eleven times the diameter of the Earth, but rotates faster than any other planet, its day being slightly less than 10 hours in length. It rotates so fast, indeed, that the equator is much bulged out by the centrifugal force. Since Jupiter is only a third denser than water we are not surprised to find that it possesses an enormously thick atmosphere, how thick we do not know. Ammonia, methane (marsh gas), and hydrogen are known to be present above the gigantic clouds that we can see from the outside (Fig. 8). These markings are certainly clouds because their forms are ever changing. Their general structure is banded parallel to the equator, as though the clouds were being blown along by "trade winds" that result from the rapid rotation. Mysterious radio noise from Jupiter indicates that the interior rotates as a solid.

The atmospheres of the other giant planets are very similar to that of Jupiter, the differences being attributable in large measure to the fact that the planets farther from the Sun are colder at their surfaces. Neptune, 30 A.U. from the Sun, is a frigid world by

Fig. 8. Jupiter, photographed with the 200-inch reflector in blue light, showing the Great Red Spot. (Photograph by the Mount Wilson and Palomar Observatories.)

our standards because it receives only 1/900 as much heat and light from the Sun as we receive. Solid carbon dioxide ("dry ice") near its melting point is hot compared to the probable temperature at the surface of Neptune, well below $-300°$ F. Nitrogen gas would be frozen, likewise oxygen.

Although the giant planets are cold and uninhabitable, their great masses and wide separation in space allow them to control astonishingly large families of satellites. Jupiter is again first, with twelve moons, Saturn is second with ten (? see p. 188), while Uranus has five and Neptune only two (Figs. 9 and 10). The brightest of Jupiter's family, Ganymede, is larger than Mercury, while another, Callisto, and Saturn's Titan are about as large. Neptune's Triton and two of Jupiter's satellites are comparable in size to our Moon, while the others range in diameter from that of small asteroids to about half that of the Moon. The systems of Jupiter and Saturn are really miniature solar systems in every respect except that the primary planets do not send out light by themselves but shine by reflected sunlight only. The great planets are more massive when compared to their largest satellites than is the Sun when compared to Jupiter or Saturn.

The similarity with the whole solar system is even more striking

Fig. 9. Neptune and Triton. (Photograph by the Lick Observatory.)

Fig. 10. Saturn and four of its satellites: Titan (*top*), Rhea, Dione, and Tethys. The rings and disk are "burned out" in this photograph of the satellites, made with the 82-inch McDonald reflector. (Photograph by the Yerkes Observatory.)

in the system of Saturn because this planet not only controls nine satellites, equal to the number of known planets about the Sun, but also possesses a family of miniature asteroids, which comprise the great rings (Fig. 11). These rings are so close to Saturn itself that in the poor early telescopes they looked like ears or appendages. Galileo, who was the first person to see Jupiter's four bright satellites, sometimes drew Saturn as consisting of three pieces—a central body with symmetric side sections (Fig. 12). We know now that the rings are made of small particles possibly covered with ice and revolving about Saturn in a plane that is relatively thinner with respect to its width than a sheet of paper. When seen at different angles they present different appearances, from invisibility, if seen edge on, to wide rings at the greatest angle possible.

On beyond Neptune lies the more recently discovered Pluto. Little is known about Pluto as a planet. Our best information suggests that it is much smaller and less massive than the Earth; it rotates in 6 days 9 hours. Probably there is no atmosphere.

To complete this quick introduction to the solar family, some mention must be made of the remarkable comets (Fig. 13). These strange wanderers have excited more superstitious fear in the

Fig. 11. Saturn, the ringed planet. (Photograph by the Mount Wilson and Palomar Observatories.)

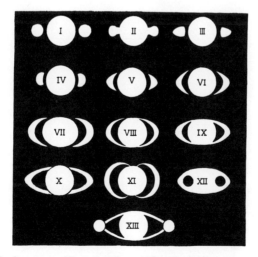

Fig. 12. Early drawings of Saturn. (From T. E. R. Phillips and W. H. Steavenson, eds., *Splendour of the Heavens,* 1923; courtesy of Hutchinson and Co., London.)

Fig. 13. Head of Comet Whipple–Bernasconi–Kulin, February 28, 1943. (Photograph by H. Giclas, Lowell Observatory.)

human mind than any other class of celestial bodies. Most comets move in exceedingly elongated orbits, approaching the Sun for only a short time each revolution. When distant from the Sun they become too faint for observation, but they brighten enormously at perihelion, when closest to the Sun. At this time they become so active that they waste an appreciable part of their substance into space, and produce a great coma of gases and small dust particles about their nuclei (see Fig. 13). Sunlight and gases blown out from the sun force the cometary gases and dust back from the head in a great tail, sometimes a multiple tail, and always complicated in structure. The structure and brilliancy of the tails are obvious from inspection of Fig. 14. Most astronomers now accept the author's

Fig. 14. Two photographs of Halley's comet. The telescope was made to follow the comet's motion during a time exposure and hence the star images are trailed. (Photographs by the Lick Observatory.)

theory that comets are fundamentally balls of ice and dirt and that they become active only when so close to the sun that the ice vaporizes.

In later chapters we shall become much better acquainted with each of the members of the solar family met in the present chapter. Each has a character more appreciated on closer contact. Also there are many provocative problems of structure and origin and even some family skeletons. In the next chapter we shall look into the important problem of family unity, the binding force that keeps each member in its place.

2

How the System
Holds Together

A mighty and all-pervading force enables the Sun to hold the
planets in their orbits of revolution, and empowers the planets
to retain their satellites. The discovery of the law of universal gravitation, which describes this force, stands as a monumental feat of the
human mind. Only a genius like Sir Isaac Newton (1643–1727)
could have started with the observational material and theories of
his day, developed a new form of mathematics to solve the dynami-
cal problems, and finally welded the observations and mathematical
theory to form a simple yet universal law. A better understanding
of his achievement can be obtained by glancing backward at the
scientific foundation from which he started.

   During the two centuries preceding Newton's activities, a few
European scientists had been amassing evidence and arguments to
disprove the concept that the universe is centered on the Earth,
benevolently lighted by the Sun with the Moon, planets, and stars
as cheerful decoration. Nicholas Copernicus (1473–1543) is credited
with the major efforts in bringing into disrepute the idea of a fixed

Earth, an idea long cherished by the followers of Aristotle, the ancient Greek philosopher (384–322 B.C.). Actually many of the ancient Greeks favored the philosophic concept of a moving earth, but Aristotle's opinion carried great weight. An early impression of planetary motions as seen from a fixed Earth is shown in Fig. 15. Note that this system of Ptolemy's (2nd century A.D.) represented the known facts extremely well, and simply.

Once the concept of a moving Earth was recognized as likely, although not well proved, the subsequent task of finding out how the Earth moves and why it moves was still difficult. The stars, actually distant suns, are too far away to indicate by their yearly motion the 93-million-mile swing of the Earth about the Sun, even by measures made long after the invention of the telescope. One can very well sympathize with the critics of the new theory who stoutly maintained that solid earth was solid earth, obviously fixed in space. "If it were moving, as those young upstarts would have one believe, why do not the stars swing back and forth across the sky during the year?" The argument was absolutely sound, and was disproved only during the nineteenth century by means of the most accurate observing techniques. The nearest star, Proxima Centauri, is 270,000 astronomical units distant. From it the Earth's orbit would appear to have a radius smaller than the diameter of a human hair as seen 15 yards away from the eye. Thus the yearly oscillation of Proxima Centauri is an angle less than the apparent

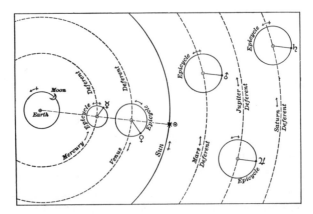

Fig. 15. The Ptolemaic system. According to the ancient Greco-Egyptian astronomer Ptolemy, the planets moved in small circles about fictitious planets that moved in large circles about the fixed Earth. (From C. A. Young, *A Text-Book of General Astronomy*, 1888; courtesy of Ginn and Co.)

Fig. 16. The Earth turns on its axis, as shown by this 8-hour exposure with a fixed camera pointing at the North Pole. The heavy trail near the center was made by Polaris. (Photograph by Fred Chappell, Lick Observatory.)

motion of the hair if moved through twice its diameter. For all other stars the annual motion is even smaller.

While the argument was raging about the motion of the Earth, the difficulty of predicting future positions for the Sun and planets, within the increasing degree of accuracy to which they could be observed, was becoming more and more serious. The invention of the clock accentuated the need for better predictions and more effective instruments to measure directions on the sky. It was necessary to know accurately how the planets actually move through space. The daily rotation of the Earth (see Fig. 16) and its yearly revolution, as we know today, complicate the problem enormously, because the observations must all be made from the Earth, a body itself in motion. In addition, light rays must pass through the atmosphere, which can bend them as much as half a degree when near the horizon.

The effects of rotation and atmosphere can largely be removed if one establishes the relative positions of the stars in a fixed system covering the entire sky and then measures the positions of the planets with respect to the stars. The apparent motion of Mars during one *opposition* (see Appendix 2 for definition of planetary configurations) from the Sun is shown in Fig. 17. This peculiar curve on the star background little resembles the smooth curve of the actual space motion already shown in Fig. 1.

In the sixteenth century the great Danish astronomer, Tycho Brahe (1546–1601), set about doing what he could to improve the knowledge of planetary motions. His principle of action is one that should always be remembered by every scientist, because it embodies the very essence of good science. Tycho Brahe made the best instruments he could, made the best observations possible with them, and then carefully studied his instruments to determine the size of the errors that should be expected. His long series of observations of Mars were minutely analyzed by Johannes Kepler of Württemberg (1571–1630), who experimented with every kind of motion that he could devise for the planet. Some types of eccentric motions of Mars about the Sun would fit the observations *almost* as well as they

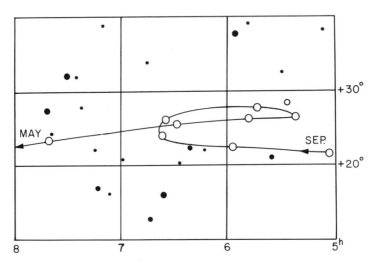

Fig. 17. Mars followed this path against the background of stars for 8 months. The circles represent positions at month intervals.

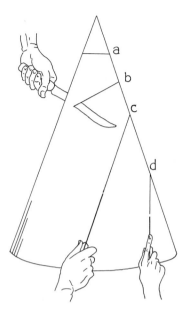

Fig. 18. A cone is sectioned by a plane to produce (*a*) a circle, (*b*) an ellipse, (*c*) a parabola, or (*d*) a hyperbola. These plane curves are called conic sections.

should, but Kepler was obsessed by scientific ideals. His perseverance led him finally to discover three very simple laws describing the motion of a planet about the Sun. A simple law, if it fits the observations well, always pleases and encourages a scientist. Kepler was certain that he had learned the truth about planetary motions, and time has corroborated his opinion. His laws also describe the motions of space vehicles when they move unpowered between the planets as well as the motions of satellites about planets.

Kepler's first law states that *the orbit of a planet is an ellipse with the Sun at one focus.* Now an ellipse is one of the simplest closed curves on a plane, one that has always delighted mathematicians because of the many simple theorems to which it is susceptible. To obtain an ellipse is nearly as easy as to draw a circle. Simply take a cone (right circular, if a mathematician is nearby) and slice it with a plane. The curve where the two surfaces intersect is an ellipse, as in (*b*), Fig. 18. You may, of course, be ingenious and pass the plane through the vertex to obtain only a point, or perpendicular to the axis to obtain a circle (*a*), or parallel to a side. In the last mentioned case the ellipse never closes, becoming a parabola (*c*), or even a hyperbola (*d*), if the plane is more nearly vertical. These possibil-

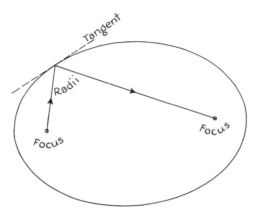

Fig. 19. An ellipse. Radii from the two foci make equal angles with a tangent. The eccentricity of this ellipse is 0.72.

ities are no problem to the mathematician, who calls all the possible curves *conic sections* and then proceeds to derive even more general theorems about them.

There are a number of theorems about the focus of an ellipse. For example, if we draw a line from one focus and reflect the line from the ellipse at an equal angle to the tangent, the line will always pass through the second focus, as in Fig. 19. The situation is even simpler for a parabola where rays from the focus are reflected as a parallel bundle. This is the principle of a searchlight or an automobile headlight. In reverse, it is the principle of the reflecting telescope where parallel light rays from a distant star are brought to a point focus by reflection from the surface of a parabolic mirror (Fig. 20).

Another noteworthy property of an ellipse is that from any point of the ellipse the sum of the distances to the two foci is constant. This property suggests a very easy method of drawing an ellipse. Stick two strong pins into a sheet of paper at the points where the foci are to be. Then place a closed loop of string about the two pins, stretch the loop taut with the point of a pencil, and draw the ellipse by swinging the pencil around the pins inside the taut loop (Fig. 21). If the two pins are together one draws a circle, the simplest ellipse.

According to Kepler's first law, the Sun is always at one focus of the ellipse, the other focus being empty—a completely neglected mathematical point. Various possible orbits are drawn in Fig. 22. The point nearest to the Sun is *perihelion* and the point farthest away is *aphelion*. The *mean distance* is half the sum of the perihelion and

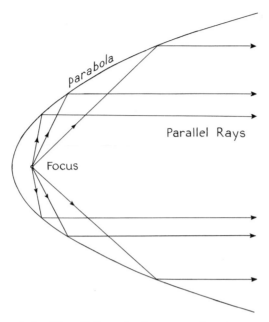

Fig. 20. A parabola. All radii from the focus are reflected into parallelism by a parabola.

aphelion distances, or the semimajor axis of the ellipse. The shape of the orbit is measured by the *eccentricity,* which is the difference of the aphelion and perihelion distances divided by their sum. The eccentricity is zero for a circle, 1.0 for a parabola, and about 0.5 for a man's hatband.

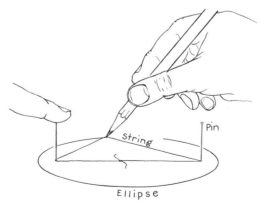

Fig. 21. Drawing an ellipse by means of two pins and a loop of string. This method works well except for the knot. It is better to tie the string at one pin.

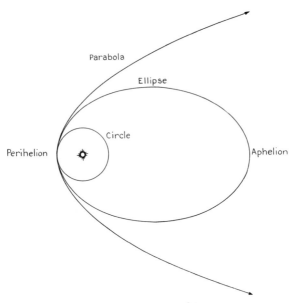

Fig. 22. Various orbits about the Sun. Comets follow orbits that are very elongated ellipses, nearly parabolic close to the Sun. Planetary orbits are ellipses but nearly circular.

The Earth's orbit is almost a circle, having an eccentricity of only $\frac{1}{60}$. To the eye such an ellipse resembles a circle fairly well drawn, but the focus is clearly not in the center. Mercury and Pluto are the only planets with orbits that deviate much from circles, their eccentricities being 0.21 and 0.25, respectively. The distance of Pluto from the Sun thus varies from 30 A.U. at perihelion, just less than Neptune's mean distance, up to 50 A.U. at aphelion. One checks this calculation by noting that the mean of 30 and 50 is 40, the mean distance of Pluto in astronomical units, and noting that $(50 - 30)$ divided by $(50 + 30)$ is $\frac{20}{80}$ or 0.25, the eccentricity.

Kepler's second law of planetary motions is simpler than the first. It states that *a line joining a planet to the Sun sweeps out equal areas in equal intervals of time.* Accordingly, when a planet is near the Sun at perihelion it must move at a greater speed than when it is farther away, say at aphelion, as is shown in Fig. 23. For Pluto the speed is 3.8 miles per second at perihelion and 2.3 miles per second at aphelion. The ratio is 5/3, as we might have guessed from the ratio of the two distances. At perihelion (about January 1), the Earth

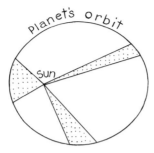

Fig. 23. Equal areas are swept out in equal intervals of time. A planet moving in this orbit would pass through each of the shaded areas in the same time. The eccentricity of the ellipse is 0.5.

has increased its speed by 0.6 miles per second from its aphelion speed of 18.2 miles per second.

Kepler's third law, the harmonic law, states that *the squares of the periods of revolution of the planets about the Sun are in the same ratio as the cubes of their mean distances.* This law provides an easy method for calculating the period if one knows the mean distance of a body revolving about the Sun. Express the mean distance in astronomical units. Cube this distance. The square root of the cube is the period of revolution in years. For the Earth the formula checks from the definition of the astronomical unit and the year: the square root of 1^3 is 1; the period of the Earth is 1 year. For Neptune the mean distance is 30 A.U., and $30^3 = 27,000$. The square root of 27,000 is 164, Neptune's period in years. More accurately the period is 164.8 years, obtained by using a more precise value of the mean distance.

By means of Kepler's three laws, the elliptic law, the law of areas, and the harmonic law, a planet's motion can be predicted far into the future. Only three corrections are applied today. The masses of the planets change the laws slightly; one planet will disturb the motion of another; and a slight correction must be made in the case of Mercury's orbit because of an effect predicted by Einstein's theory of relativity.

Newton was fully conversant with Kepler's laws for describing the motions of the planets and also with Galileo Galilei's (1564–1642) revolutionary idea that all bodies fall at the same rate regardless of size. Galileo had demonstrated this idea by dropping large and small weights from a tower, which may have been the Leaning Tower of Pisa. Galileo held another revolutionary idea, that in space bodies would continue to move indefinitely unless stopped by some force. But Galileo's well-known difficulty with the Church arose largely from his teaching Copernicus' theory that the earth really moves.

Newton extended Galileo's ideas about the motion of material bodies in empty space and crystallized them into three simple laws. These principles of motion are so familiar to us today that they are listed only for the sake of completeness. The first states that *a body remains at rest or maintains a uniform motion in a straight line unless acted upon by a force;* the second states that *the rate of change of motion is proportional to the force acting* (really a definition of *force*); and the third states that *action and reaction are equal but opposite in direction.* Obvious applications exist everywhere in our modern world of machines. Lack of measurement of frictional forces, both with the air and between moving parts in machinery, is the one difficulty that prevented the laws from being discovered much sooner.

With all of these principles in mind, Newton began to ponder the problems of the motions of the Moon and the planets. Since, by gravitation, the Earth attracts an apple, or a cannon ball, or a feather, each with a force proportional to the mass, why should it not attract the Moon? By all rights the Moon should move in a straight line unless it is acted upon by a force—but the Moon actually moves in a curved path about the Earth. Therefore, it must be continually falling toward the Earth, the rate of fall being measured by the deviation from motion in a straight line (Fig. 24). Thus, the attraction of the Earth must produce a force on the Moon of exactly the right magnitude to cause the Moon to fall as it does.

"But how does the force of gravity decrease with distance from the earth?" asked Newton. To answer this he first calculated the law of centripetal force, the central force exerted when a ball is whirled at the end of a string fastened at a point. To produce this centripetal force he found that gravity should fall off as the inverse square of the distance to the attracting center. Then from Kepler's laws he independently deduced that the planets must also be attracted toward the Sun by an inverse-square law of force.

Newton was then ready to check his theory by the Moon's motion. Here he ran into difficulties. He first used an erroneous value of the Earth's dimensions and also found some trouble in proving that a spherical Earth should act gravitationally as though its

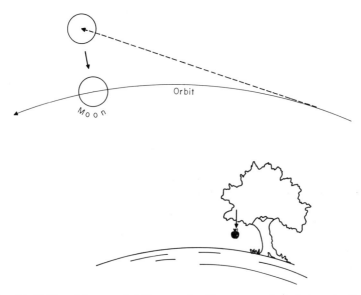

Fig. 24. Falling Moon and falling apple. Newton concluded that the Earth's gravity causes the Moon to fall toward the Earth from a straight line and the apple to fall from a tree according to the same law.

entire mass were concentrated at its center. Finally, however, the check worked. Putting all the evidence together he came to the conclusion that *every particle of matter in the universe attracts every other particle with a force that varies directly as the product of their masses and inversely as the square of the distance between them.* This law of universal gravitation accounts for all the complicated motions in the solar system to the high degree of accuracy possible in astronomical measurement (about one part in a million). The only error is a forward motion of the perihelion of Mercury, by about 43″ in a century, which is explained by a slight correction to Newton's law as predicted by Einstein's theory of relativity. An angle of 43″ would be subtended by the iris of the eye at a distance of about 50 yards.

Thus the solar system holds together by the attraction of the Sun upon the planets, and the satellite systems by the attraction of the planets upon their satellites. The problem would be very simple and completely solved by Kepler's laws if it were not for the unhappy circumstance that all of the planets attract each other, as well as their satellites and the Sun. This universal attraction complicates the problem so much that there is no exact mathematical solution. The only saving grace lies in the fact that the planets are much less massive than the Sun so that the forces of interattraction, proportional to the masses, are much smaller than the attraction of the Sun. The satellites, likewise, are much less massive than their planets. Consequently, Kepler's laws can be used to obtain an approximate solution for the motions, small corrections being made on the basis of the interattractions. These corrections are known as *perturbations,* because the other planets *perturb* the motions of the one under investigation.

The most difficult classical problem of perturbations occurs in the Earth-Moon system where the Sun perturbs the motion of the Moon about the Earth and where our observations are so excellent because the Moon is so close. Strictly speaking, the Earth perturbs the motion of the Moon about the Sun because the Sun actually attracts the Moon with a force nearly twice as great as the attraction of the Earth on the Moon. Nevertheless, there is no danger that the Sun can steal the Moon away from the Earth and leave us without inspiration on warm summer nights. The system is so compact, the Earth and Moon moving so nearly together, that the Sun's attraction serves only to keep both bodies moving about it in an average path that is elliptic. The chief results are: first, that the Moon's orbit is never convex *toward* the Sun (Fig. 25), and second, that astronomers have much more work to do in predicting the Moon's motion. One single equation for the motion of the Moon covers some 250 large-size pages and represents the major effort of a lifetime.

The fact that the Earth is not exactly a sphere (Chapter 5) and does not attract precisely from its center adds a further slight complication to the motion of the Moon. The motions of artificial

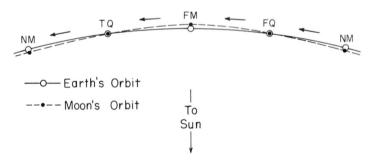

Fig. 25. The Moon's orbit about the Sun. The deviations from a perfect ellipse are greatly exaggerated in the diagram; even so, the Moon's orbit is always concave toward the Sun, as can be seen by tipping the page and sighting along the broken line.

Earth satellites, however, are much disturbed by this effect, and, surprisingly enough, even by the pressure of sunlight. Furthermore, the resistance of the Earth's high atmosphere tends to haul them down. Their periods of revolution, too, are so short—a minimum of only 88 minutes—that all these complicated effects must be calculated rapidly, sometimes in minutes. Thus electronic calculating machines, thousands of times faster than older methods of calculating, are absolutely essential to the success of our modern satellite and space-vehicle programs. Astronomers greatly prize these machines, incidentally, to help solve older problems of celestial mechanics in weeks instead of lifetimes.

Among the planets, Jupiter is definitely the solar system's bad boy in disturbing the motions of all the planets and asteroids. With a mass equal to 0.001 that of the Sun, a "lion's share" of the mass of the entire planet family, Jupiter produces by far the largest perturbations, particularly among the asteroids which are nearest to it in space (Fig. 7). If the orbit of an asteroid is calculated without allowing for Jupiter's attraction, the errors of prediction may amount to several degrees within a few years. Asteroids sometimes get "lost" in this fashion, until they are independently rediscovered and identified by their orbital path and brightness.

Jupiter also forces the asteroids to move with periods that are not related by simple fractions (such as ½ or ⅓) to the period of Jupiter (Fig. 26). In Saturn's rings, which we have mentioned as similar to the asteroid system, exactly the same effect occurs. The fact that there are *rings* rather than a single ring results from

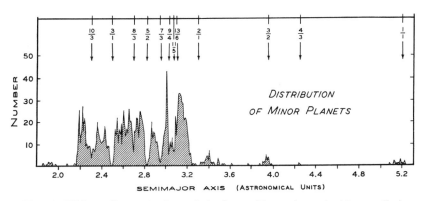

Fig. 26. Kirkwood's gaps in the periods of asteroids are shown in this compilation. The fractions are the ratios of the period of Jupiter to that of the asteroids. (Courtesy of *Sky and Telescope*.)

Fig. 27. Saturn's rings as drawn by B. Lyot. (Courtesy of A. Dollfus.)

the perturbations of the minute moonlets in the rings produced by three inner satellites, Mimas, Enceladus, and Tethys. These satellites force the small particles in the rings to avoid certain speeds in moving about Saturn, with the result that dark spaces show at the critical distances. Cassini's division (G. D. Cassini, 1625–1712), the conspicuous dark space between the outer and middle rings in Fig. 27, covers a region in which the tiny moonlets would move with periods of one-half that of Mimas, one-third that of Enceladus, or one-quarter that of Tethys, slightly more than 11 hours. Other expected "dark rings" can also be observed.

The fact, mentioned earlier, that there exists such a large ratio of mass between the Sun and the planets, and between the planets and their satellites, cannot be just a chance circumstance. The solar system would be drastically different were the planets comparable in mass with the Sun. It is important to have one body always near the *center of gravity,* the point about which all the bodies would balance if supported by weightless rods under a uniform attraction similar to gravity on Earth. No one of the planets could stay near the center of gravity, as the Sun does now in the solar system, but all would be moving in complex curves of almost unpredictable form. Although mathematics cannot provide a useful solution even for the case of three bodies of nearly equal mass, it demonstrates that in our hypothetical case there would be disastrous results. Some planets would be eliminated by collisions and others tossed out of the system until, most probably, the system would eventually consist of two of the largest bodies moving about each other at a moderate distance, with smaller companions or satellite systems. Other bodies might remain in the system at very great distances from the largest ones. Our experience with double and multiple stars indicates that they tend to occur in pairs separated by relatively large distances from other pairs in the system.

At present there is no danger that the solar system will lose any planets or that any collisions of importance will occur. Our absolute certainty that there can be no disasters covers only hundreds of millions of years, but the thought of possible disasters later on is really not very depressing at the moment.

3

The Discoveries
of Neptune and Pluto

The remarkable story of the discoveries of Neptune and Pluto begins actually with the discovery of Uranus, because without observations of Uranus the two later discoveries would have been delayed for many years. Indeed, the detection of Uranus marks the beginning of a new epoch in the history of astronomy, for Uranus was the first planet to be "discovered." Mercury, Venus, Mars, Jupiter, and Saturn have always been visible to the eye of any man who looked skyward (unless the eyes of our prehistoric ancestors were much inferior to our own).

Sir William Herschel (1738–1822; Fig. 28), perhaps the most assiduous observer of all time, first detected the small disk (3.6 seconds of arc) of Uranus in 1781. His account of the discovery shows clearly that he was not immediately certain of the true character of the new object. From the *Philosophical Transactions* of 1781 we read, "On Tuesday the 13th of March, between ten and eleven in the evening, while I was examining the small stars in the neighborhood of H Geminorum, I perceived one that appeared visibly larger than the rest; being struck with its uncommon magnitude, I compared it to H Geminorum and the small star in the quartile between Auriga and Gemini, and finding it to be much larger than either of them, suspected it to be a comet."

Herschel's announcement of the new object as a comet was natural and conservative, whatever he may have suspected concerning its true nature. Several months of observation and calculation were required to demonstrate that no cometary motion would satisfy the observations and that the "comet" could be nothing less than a new planet.

It was an extraordinary keenness of eye and judgment that enabled Herschel to distinguish the planet from the nearby stars by its appearance alone. Other observers, while measuring the positions of neighboring stars, had 17 times measured the position of Uranus and had noticed no unusual aspect. Some of the great contemporary astronomers had difficulty in identifying the planet even after they had been informed of its exact position on the sky.

Uranus did not become the planet's official name for several years. It first bore the title "Georgium Sidus" (Herschel's appellation in honor of King George the Third), and was also called "Herschel" for its discoverer. The present name was finally adopted, in conformity with the naming of the other planets.

In spite of Uranus' slow motion (with a period of 84 years), its orbit could be well determined within a relatively short time after discovery because of the 17 inadvertent observations made before Herschel noticed the disk. The first observation, made in 1690, was earlier by nearly a complete revolution of Uranus. The orbit calculators found some difficulty in reconciling all the observations, but the chance for errors in the observations, or for deviations because of the perturbations by other planets, seemed great enough to account for the discrepancies. However, when Uranus began to deviate appreciably from its computed path, even after careful allowance had been made for the perturbations by Jupiter and Saturn, several astronomers began to suspect that an unknown planet might be disturbing the motion of Uranus.

In the second and third decades of the nineteenth century the deviations were large enough to arouse suspicion, but the mathematical difficulties in predicting the position of the unknown planet seemed insurmountable at that time. By the year 1845 Uranus had moved out of place by the "intolerable quantity" of 2 minutes of arc, an angle barely resolvable by the naked eye. Urbain Jean Joseph

Fig. 28. Sir William Herschel.

Fig. 29. Percival Lowell, whose prediction and enthusiasm led eventually to the discovery of Pluto.

Leverrier (1811–1877), the great French astronomer, showed in 1846 that no possible orbit for Uranus could reconcile all the observations within their reasonable errors. He concluded that the deviations could be explained only by the hypothesis of an unknown massive planet beyond the orbit of Uranus.

Later in the year 1846 Leverrier completed his calculations for the position of the hypothetical planet, and was so confident of his analysis that he dared to predict its position and that it would show a recognizable disk. He sent his predictions to the young German astronomer J. G. Galle (1812–1910), who discovered the actual planet *on the same night* that he received the prediction. Neptune's position on the sky lay within 1°—less than two Moon diameters—of the position forecast by Leverrier. Galle's immediate success was due to his access in Berlin to a new star chart of the appropriate sky region. A quick telescopic survey showed the new object where no star had been seen before. A close inspection verified the existence of a disk, too small to be distinguished easily.

This remarkable discovery of a new planet by means of mathematical deduction is a landmark in the history of astronomy. Like many great discoveries, it must be credited to more than one man. While Leverrier had been making his brilliant calculations, a young and unknown English mathematician, J. C. Adams (1819–1892), arrived independently at the same result by a somewhat different method. Adams's calculations had, indeed, been completed some eight months before Leverrier's, but unhappy circumstances prevented the English observers from anticipating Galle's discovery. The Berlin star chart that had so materially assisted Galle was not then available in England. Hence the astronomer James Challis (1804–1883), at Cambridge half-heartedly began searching for the planet by the arduous method of plotting all the stars in the region, with the intent of reobserving them later in order to detect the planet by its motion. Neptune might also have been found at the Royal Observatory in Greenwich except that the Astronomer Royal, G. B. Airy, held a negative attitude toward theory and simply did not believe that Adams could make a valid prediction.

The entire turn of events was heartbreaking for Adams, who had apparently planned his investigation some years before he had the opportunity to execute it. Posthumously the following note was found among his effects: "1841, July 3. Formed a design, in the beginning of this week, of investigating, as soon as possible after taking my degree, the irregularities in the motion of Uranus, which were as yet unaccounted for: in order to find whether they may be attributed to the action of an undiscovered planet beyond it, and if possible thence to determine the elements of its orbit, etc., approximately, which would probably lead to its discovery."

It is a pleasure to record that both Leverrier and Adams now share equally the honor of having predicted the existence and position of Neptune. Galle, of course, receives his full credit for actually having found it. Like Uranus, Neptune had been mistaken for a star in the course of previous measurements of stellar positions.

The conquest of the solar system by Newton's law of gravitation and by painstaking observation has been continued in the present century. Efforts have culminated in the discovery of Pluto, under circumstances surprisingly similar to those related for Neptune. Again, an early search actually included the new planet but perverse fortune prevented its detection until much later.

At the beginning of this century Percival Lowell (1855–1916), who founded an observatory at Flagstaff, Arizona, for the purpose of observing the planets, particularly Mars, became actively interested in a possible trans-Neptunian planet. He reinvestigated the orbit of Uranus and concluded that the apparent errors of observation could be materially reduced by the inclusion of perturbations by an unknown planet. His calculations of the orbit and positions of Planet X were not published until 1914, although his search for the planet was begun in 1905. Twenty-four years later, in 1929, a new 13-inch refracting telescope to expedite the search was completed and was installed at the Lowell Observatory.

A young assistant, Clyde Tombaugh, was assigned the task of systematically photographing regions of the sky along the ecliptic. For each region he made two long-exposure photographs, separated in time by 2 or 3 days. Then, in search of the predicted planet, he very carefully compared the resulting photographic plates. Comparisons were made by means of a *blink comparator,* a double-microscope apparatus that enables the observer to inspect the same area of the sky on two plates alternately (Fig. 30). Any object that has moved on the sky during the interval between the two exposures appears to jump back and forth among the stars, which appear to remain fixed.

On March 12, 1930, less than a year after the institution of its new observing program, the Lowell Observatory by way of the Harvard bureau telegraphed astronomical observatories the follow-

Fig. 30. Clyde W. Tombaugh at the blink comparator where he sat for 7000 hours in his planetary search. (Photograph by the Lowell Observatory.)

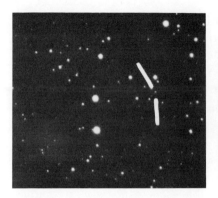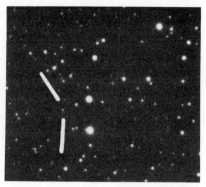

Fig. 31. Discovery photographs of Pluto: (*left*) January 23, 1930, (*right*) January 29, 1930. (Photographs by C. W. Tombaugh with the 13-inch Lawrence Lowell Refractor of the Lowell Observatory.)

ing announcement: "Systematic search begun years ago supplementing Lowell's investigations for Trans Neptunian planet has revealed object which since seven weeks has in rate of motion and path consistently conformed to Trans Neptunian body at approximate distance he assigned. Fifteenth magnitude. Position March twelve days three hours GMT was seven seconds West from Delta Geminorum, agreeing with Lowell's predicted longitude."

The astronomical world soon unanimously adopted the name Pluto as appropriate to this planet, which moves in the outer regions of darkness. The first two letters of the name are, moreover, the initials of Percival Lowell, who had died in 1916, only 2 years after the publication of his detailed prediction.

Subsequent orbits, based on prediscovery photographs of the new planet, show that it moves about the Sun with a period of 249.9 years, in an orbit inclined 17° to the mean plane of the other planets. At perihelion the orbit passes within that of Neptune, but because of the high inclination the two bodies cannot collide.

Only the perversity of chance kept the discovery of Pluto from being made by the Mount Wilson astronomers in 1919. At that time Milton Humason, at the request of William H. Pickering (1858–1938), who had independently made calculations of the assumed planet's position, photographed the regions around the predicted position and actually registered the planet on some of the plates. Pluto's image on one of the two best plates, however, fell directly upon a small flaw in the emulsion—at first glance it seemed to be a part of the flaw—while on the other plate the image was partly superimposed upon that of a star! Even in 1930, when the 1919 position was rather well known from the orbit, it was difficult to identify the images that had been produced by Pluto 11 years before.

Unless Pluto is fantastically dense, or an extraordinarily poor reflector of light, it cannot be massive enough to have produced the deviations in planetary motion on which its prediction was based. Thus, most astronomers believe today that the prediction was fortuitous. Nevertheless, the discovery, following the relentless search for the planet, represents a crowning achievement in scientific progress. All of the members of the Lowell Observatory staff deserve the highest praise for the painstaking work and the consequent result.

Tombaugh has continued the Lowell Observatory search, covering the entire sky, but finds that there exist no more planets within the discovery range of the 13-inch telescope. If other planets exist they must be considerably fainter than Pluto, which means that they must be either farther away or smaller. To continue the quest for much fainter planets with one of the large telescopes, say the 200-inch reflector at Mount Palomar, would be impractical. The larger telescopes progressively photograph smaller areas of the sky. A search of the entire sky to the limiting brightness attainable with the 200-inch telescope would require its continuous use every hour of every clear moonless night for centuries. The discovery of possible planets beyond Pluto will be very difficult, unless luck plays a role or unless new observing techniques are applied. The distances are hopelessly too great for radar telescopes. A large optical telescope in space or on the Moon, coupled with television techniques and with automatic search equipment, might conceivably succeed. Some astronomers doubt that there are more sizable planets to be found.

4

Weights and Measures

Weighing the planets and finding the distances between them are naturally most important in learning about their true character. Only from a knowledge of the masses can we begin to ascertain the real structure of the individual bodies. Furthermore, the landing of space vehicles on planets and satellites demands the utmost precision in distances, dimensions, and masses. Thus to increase our knowledge, we place demands on the accumulated fund of knowledge.

The Distance to the Sun

The stars, apparently bright points scattered all around the sky, form a magnificent reference system against which we can measure the directions of moving bodies in the solar system. As we noted in Chapter 2, the stars are so distant that our motion does not affect their directions perceptibly, except in a few cases. Thus, by measuring planetary and solar directions precisely with respect to the stars, we determine the directions of the Sun as well as the planets with respect to a well-defined reference system.

With accurate observations and calculations now spread over hundreds of years, we can apply Newton's law of universal gravitation and calculate all the relative directions and distances to the Sun and planets with a precision of about one part in a million. But all these accurate distances are determined in terms of the *astronomical unit,* half the length of the Earth's ellipse about the Sun, not in terms of feet or miles. For purposes of prediction, the arbitrary nature of this unit of distance makes practically no difference, but no scientist relishes a measuring rod whose length is unknown. Also, in maneuvering space vehicles, we need to know the actual distances in miles.

In measuring the astronomical unit we are faced with the fact that the largest available yardstick is the Earth itself; its dimensions are now known rather accurately. But the radius of the Earth is less than 1/20,000 of the astronomical unit; from the Sun it subtends an angle of only 8.8 seconds of arc, the *geocentric parallax* of the Sun (see Fig. 32). Although the Sun's parallax can be measured by simultaneous observations at two widely separated stations, the angle is so small that the percentage error becomes large; hence no precise determination of the length of the astronomical unit can be made by measurements of the Sun.

A better method is to measure in miles the distance to some body that comes close to the Earth. Since the body's distance in astronomical units at any time is determinable from many observations and calculations, we can compare the two values to find the number of miles in an astronomical unit. The Moon, however, will not serve for this purpose because its distance cannot be calculated in astronomical units without introducing the Earth's mass. As Newton found (Chapter 2), the Moon's motion is primarily a measure of the Earth's gravity. Mars, under the most favorable conditions,

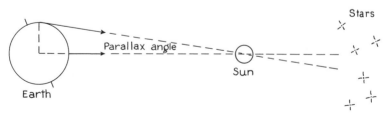

Fig. 32. The solar parallax is the angle subtended by the Earth's radius as seen from the Sun. The geocentric parallax of any celestial object is the corresponding angle from that object. See Fig. 50 for stellar parallax.

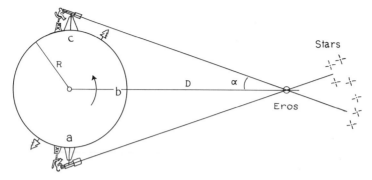

Fig. 33. Measuring the distance to Eros. The observer simply allows the Earth to move him from *a* to *c* in order that he may measure the parallax angle, α, at Eros and calculate the distance, *D*.

comes within only 34,600,000 miles of the Earth. Venus comes closer, 26,000,000 miles, but at that time it is nearly in the direction of the Sun and cannot be well measured (refer to Fig. 4) except by radar, as we shall see. In the classical astronomical method, however, the best object for the purpose turns out to be an asteroid, Eros.

It is a real pleasure to find a use for an asteroid because asteroids are generally more of a nuisance than a help in astronomy. The late Walter Baade of Mount Wilson and Palomar Observatories once called them "the vermin of the skies." When Eros comes within 14,000,000 miles of the Earth, it gives a parallax of about seven times that of the Sun, or over a minute of arc. Furthermore, the asteroid looks and photographs like a star or a point of light; the resultant measurements are easy and very accurate. Observers, instead of choosing different parts of the Earth for their observations, can work independently. Each simply waits for the Earth to turn (Fig. 33) and photographs the asteroid in the evening (*b*), at midnight (*b*), and in the morning (*c*). The position of Eros among the stars changes because of the difference in the position of the observer with respect to the Earth. Knowing the instant of each photographic exposure and his exact position on the Earth, the observer can calculate the distance, *D*, to Eros in miles just as accurately as though simultaneous observations were made at different stations on the Earth. The process is similar to that of estimating a small distance with only one eye by turning one's head, or by moving it from side to side.

The motion of Eros across the sky between the observations necessitates a major correction. When this correction and other minor ones are applied, the distance can be well determined. In January 1931, Eros made one of its very close approaches to the Earth at a minimum distance of 16,200,000 miles. International cooperation among the world's leading observatories led to an improved value of the solar parallax and the astronomical unit.

In actuality, however, E. Rabe found that he could determine the solar parallax better from the motion of Eros than from the trigonometry of its distance. He used the amount that the Earth perturbed Eros as a measure of the Earth's gravity compared with the solar attraction, at distances both known in astronomical units. This gives the mass ratio of the Earth to the Sun, which can be combined with other gravity measures to calibrate the length of the astronomical unit. Eros will come close to the Earth again in 1975, but in the meantime we have a far better and more direct method for measuring the astronomical unit—radar.

When R. Price and his colleagues at the Lincoln Laboratories of the Massachusetts Institute of Technology pioneered in bouncing radar echoes from Venus in 1959 (Fig. 34), they initiated a new era in the measurement of planetary distances. The time that a radio wave requires in passage from the earth to Venus and back, coupled with the measured velocity of light in vacuum (186,282 miles per second), gives a direct measure of the distance in miles. Comparison with the calculated distance in astronomical units calibrates the latter. During the close approach of Venus in 1961 the radar astronomers at the Lincoln Laboratories, at the California Institute of Technology, and in the U.S.S.R. independently calibrated the astronomical unit as 92,956,000 miles, good to some 100's of miles. Slight uncertainties occur in the velocity of light, in the application of orbital theory to Venus' distance, and possibly in the propagation of radio waves. The corresponding solar parallax is 8.7942 seconds of arc, much more accurate than older values.

Now we know the distances in the solar system accurately enough that we can at least *aim* planetary space probes to reach planets. We know the distance to the Moon with an uncertainty of much less than 1 mile.

Weighing the Earth

The first step in weighing the planets is to weigh the Earth. The ancient Greek mathematician, Archimedes (of Syracuse, 287–212 B.C.), said that if he had a place to stand he could move the Earth.

Fig. 34. The Millstone radar antenna, 84 feet in diameter, that first bounced radio waves off Venus. (Photograph by the Massachusetts Institute of Technology Lincoln Laboratory.)

He could equally well have weighed it by observing how easily it moved when he pulled the levers. We are actually interested, not in the weight of the Earth, but in its *mass*. The weight of a body is only a measure of the Earth's attraction for it, while its mass represents the quantity of matter that it contains. One of Newton's great discoveries was that mass and weight are proportional. If we go back to Newton's laws of motion we find that mass is the measure of the force necessary to change the motion of a body by a certain amount. It takes more force to accelerate a 10-ton truck than a baby carriage at a given rate because of the difference in mass. In empty space away from attracting bodies, neither the truck nor the baby carriage would have any weight but their masses would remain unchanged.

We know just how much force the Earth exerts on a unit mass through gravitation. This force is the surface gravity, which holds us down and enables us to *weigh* things. Since gravity is proportional to the Earth's mass, the only unknown is the constant of attraction between two masses, that is, the *constant of gravity*. (For a more precise definition, see Appendix 3.)

One method of finding this constant is by measuring the attraction of a mountain on a plumb line. As in Fig. 35, the plumb line does not point straight up but points away from the mountain, because the bob is attracted toward it. We measure the force exerted on the bob by the mountain and estimate the mass of the mountain by measuring its size and composition. Since the distance to the mountain is measurable, we can calculate the constant of gravity and hence the mass of the earth.

The mountain method is fairly good but less accurate than laboratory methods. With exceedingly delicate instruments the attraction of a large ball of lead upon a smaller ball can be measured directly, giving a value for the constant of gravity. Since the weight of the small ball measures the Earth's attraction for it, the mass of the Earth is then determined in terms of the mass of the large ball by means of the inverse-square law of the attractions. If the Earth could be put on scales at the surface, it would weigh 6,600,000,000,000,000,000,000 tons or 1.32×10^{25} pounds.

The force of gravitation is an exceedingly trivial force in magnitude, becoming appreciable only when huge quantities of matter are involved. Suppose a ball were made of all the gold that has been mined in the world, say 30,000 tons; the ball would be about 46 feet in diameter. If it were placed in space, away from other

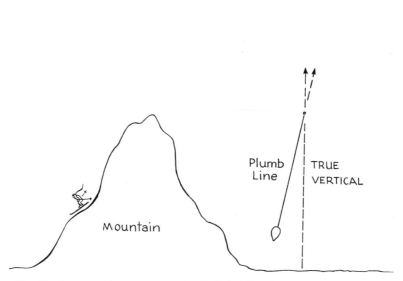

Fig. 35. A mountain attracts a plumb bob. The direction of the plumb line deviates from the vertical, to provide a measure of the mountain's gravitational force.

attracting forces, a 200-pound man sitting on it would weigh the equivalent of 0.01 ounce on the Earth. A cricket could easily lift him and a frog could kick him completely away from the ball of gold. Since men are not usually so easily diverted from gold we may conclude, in the manner of Aesop, that the force of avarice greatly exceeds that of gravitation.

The Massive Sun

Knowing the mass of the Earth, we can now calculate the mass of the Sun. The Earth continually falls toward the Sun, away from the straight line that it would follow if there were no gravitational attraction. The amount of fall is about ⅑ inch every second, in which time the Earth moves forward about 18.5 miles. The mass of the Sun required to make the Earth fall at this rate is 332,950 times the mass of the Earth, or some 4.38×10^{30} (4,380,000,000,-000,000,000,000,000,000,000) pounds.

With this knowledge of the mass of the Sun we may deduce some interesting information about its constitution. The average density is only 1.41 times that of water, while the Earth's density is 5.5, equivalent to a mixture of rock and metals. At the surface of the Sun the force of gravity is 28 times as great as on the Earth. A 200-pound man would weigh nearly 3 tons there, except that he would evaporate instantly at the temperature of 10,000°F. From only three known quantities—mass, diameter, and surface temperature—it is possible to prove that the Sun is a *gas* throughout. The temperature at the center must be about 27,000,000°F, with a density of 100 times that of water, to provide enough pressure to keep the outer gaseous layers from collapsing. No element can be a solid or a liquid in any part of the Sun; even tungsten, used in electric-light filaments, would evaporate on the surface, which is relatively cool.

A Planet with a Satellite

For a planet with a satellite the method of determining the mass is like that used in finding the Sun's mass. The attraction of the planet must always exactly equal the centripetal force, which measures the rate at which the satellite falls toward the planet to remain in its orbit. With a knowledge of this attraction, the distance of the satellite, and the constant of gravitation, we calculate the mass of the planet. We knew the mass of Neptune, 2700 million miles away, as accurately as we knew the mass of the Moon, distant only 239,000 miles, at least before radar.

Weighing the Moon

The mass of a satellite is difficult to determine because it is generally so small compared with the mass of the primary planet. The effect of the Earth on the Moon's motion is easily measured, but the Moon is so small in mass that it affects the Earth's motion only slightly. The center of the Earth moves about their common center of gravity in a very small orbit, identical in shape with that of the Moon. If Earth and Moon could be joined by a weightless rod and the rod balanced on a knife-edge under a constant gravity, the knife-edge could support the rod at their center of gravity (Fig. 36). It is this point that moves in a smooth elliptic orbit about the Sun. By carefully observing the distance of the center of gravity from the center of the Earth, we can measure the mass of the Moon (Fig. 37). The ratio of the radius of the little orbit of the Earth's center to that of the larger orbit of the Moon is the ratio of the Moon's mass to that

Fig. 36. Center of gravity is the point of balance. The figure is accurately drawn for two balls of equal density connected by a weightless rod.

of the Earth. Subsatellites moving about the Moon can, of course, provide a better measure of its mass, as can radar.

The center of gravity is about 2900 miles from the center of the Earth, so that the mass of the Moon is only 1/81.30 (or roughly 2900/236,000) times the mass of the Earth. With such a small mass, only 81,000,000,000,000,000,000 tons, the Moon is as tiny a part of the entire solar system as one drop of water in a 50-gallon barrel, or as the proverbial fly on a cartwheel. The Earth is merely 81 times as important—except to us.

Other Satellites

Only a few of the larger satellites of Jupiter and Saturn produce sufficient gravitational effects for their masses to be determined. Jupiter's four bright satellites and Titan of Saturn's system are compa-

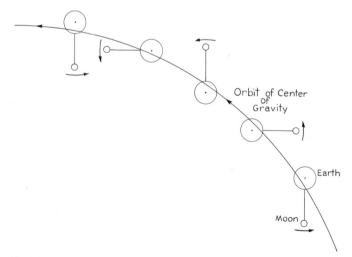

Fig. 37. Motion of the Earth about the center of gravity of the Earth and the Moon. The center of gravity moves in a smooth ellipse about the Sun. The dimensions are exaggerated, but the relative positions of the Earth's center are correct.

rable to the Moon. The other satellites are generally much less massive except for Neptune's Triton, which may be much like the Moon, although its mass is difficult to determine.

The Moon has about 3.3 times the density of water, as though made of ordinary rock. Two of the bright satellites of Jupiter are less dense and two are denser than the Moon. The third, Ganymede, is the largest satellite in the solar system, greater even than Mercury in diameter (3480 *vs.* 3025 miles) and the most massive, twice the Moon. Its density, however, is only about twice that of water, too small for rocky material. Callisto (IV of Jupiter) and Titan are much like Ganymede, but a bit smaller. The relative masses of the various

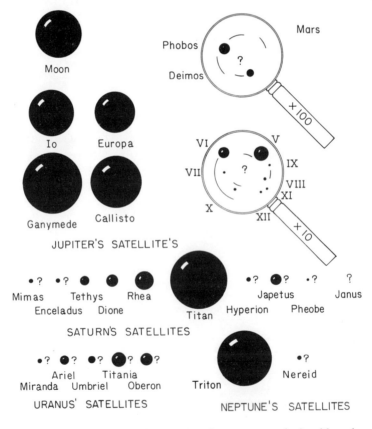

Fig. 38. The masses of the satellites in the solar system are depicted by spheres of equal density. Uncertain masses are designated by a question mark. The magnifications are in terms of diameters. The satellites are placed in the order of their distances from their respective primaries.

satellites are represented by spheres of equal density in Fig. 38. For those whose masses are not known, the measured diameters are given instead, and the satellites are designated by question marks. Diameters that were estimated from brightness alone are similarly designated. The range in mass is so great that it was necessary to magnify some of the diameters by a factor of 10 or 100 to make them visible on the diagram.

Deimos and Phobos, the Lilliputian satellites of Mars, are seen to be pitiful featherweights as compared to the average satellites. The surface gravities of these small satellites are trifling. A man on one of them would weigh only a few ounces, and would be capable of jumping several miles, if not completely off the satellite. His greatest difficulty would be to move freely without hopping upward for miles into space. The suggestion that Phobos might be an artificial satellite can be dismissed as wishful thinking.

The Masses of the Planets

The masses of planets without satellites have been calculated classically only from their observed perturbations on the motions of the other planets, by a method similar to the one used in predicting the positions of Neptune and Pluto. Venus, for example, approaches the Earth, Mars, and Mercury fairly closely and produces perturbations in their motions. Similarly, the motion of Venus is perturbed by the attractions of the Earth, Mars, and Mercury. The numerical calculation of the masses of Venus and Mercury from observations of their motions and those of the Earth and Mars is exceedingly complicated. Artificial satellites or space probes about these planets provide improved values of their masses.

Even with theories based on the best observations the mass of Mercury is not very well known. Mercury is so small and so near the Sun that its effect on the other planets is barely measurable. The best analyses indicate that the mass of Mercury is only 1/18.2 times the mass of the Earth, or about four times that of the Moon. Mercury is nearly as dense as the Earth. It would make a very fine satellite for the Earth if a transfer could be effected—but then Mercury would suffer a very great loss in prestige, while the Moon would make only a puny planet.

The mass of Venus was mentioned in the first chapter; it is only slightly less than that of the Earth—82 percent. All of the other planets, with the exception of Pluto, have satellites to expedite the calculation of their masses. The case of Pluto is an especially difficult one, chiefly because accurate astronomy is so young and because the outer planets move so slowly. Neptune, whose motion we should expect to be the most affected by Pluto, has been observed through much less than one revolution about the Sun. Prediscovery observations of Neptune lead to a calculated mass about 0.9 that of the Earth for Pluto, while D. Brouwer finds that Uranus' motion suggests ¾ of an Earth mass. In fact, no reliable determination can yet be made. Pluto's mass will thus remain uncertain for some time but it is assuredly far less than the Earth's.

Figure 39 shows the masses of the planets as balls of equal density. For comparison with Fig. 38, we recall that Mercury surpasses the Moon in mass. Therefore the sequence of masses from the least satellite, Phobos, to the greatest planet, Jupiter, is fairly uniform. We might extend the sequence to the smaller bodies, such as the asteroids and meteors, without loss of uniformity, but not in the other direction. The step in diameter from Jupiter to the Sun is a factor of ten, because the Sun is a thousand times more massive.

It would unduly burden this story to describe all of the other ingenious methods that have been used in calculating or estimating the masses of planets, satellites, comets, or asteroids. Whenever two bodies are observed to come near enough together for one to perturb the motion of the other, additional information about the masses can be determined. If close approaches occur without any observed changes in motion, an *upper limit* to the masses can be deduced. When Brooks's comet, in 1886, came within the orbits of Jupiter's inner satellites, the comet's period of revolution about the Sun was changed from 29 years to 7 years, yet no change was observable in the motions of the satellites. The comet consequently must have possessed less than 0.0001 of the Earth's mass, or it would have produced measurable perturbations.

Fig. 39. The masses of the planets, represented by spheres of equal density. The Sun's diameter, on this scale, should exceed Jupiter's by a factor of ten.

In concluding these chapters that involve Newton's law of gravitation we note that the whole foundation of astronomy rests on applications of the law as do the motions of man-made vehicles in space. Outside the solar system in the far reaches of the universe, the law is still the key to the solution of many of the most important problems. Almost no calculations of mass can be made without using the property of attraction. There is, however, evidence that Newton's law is only a first approximation, that fast-moving bodies increase their mass with speed. Einstein's theory of relativity (Albert Einstein, 1879–1955) gives the correcting factors, which are just appreciable in the case of Mercury's rapid motion under the Sun's great attraction. Solely in this instance are these corrections great enough to be detectable in present-day observations of the motions of celestial bodies.

Again, over truly incredible distances in space a different correction may be necessary; the universe appears to be expanding. If far enough apart, masses may exert a repulsive force even greater than the attractive force of gravitation. Just why repulsion should occur is not clear and perhaps some other cause is at work; perhaps the universe is just expanding or has exploded. The expansion of the universe must clearly have a deep-rooted cause, however, one that carries us to the heart of the infinite.

Since there can be no absolute in truth, each new conclusion leading on to the possibility of more general ones, we can well admire the simplicity and perfection of Newton's law which applies so exactly. To make progress, however, the scientist must search for tiny imperfections in a law or theory that seems to be perfect.

5

The Earth

Our Earth seems so large, so substantial, and so much with us that we tend to forget the minor position it occupies in the solar family of planets. Only by a small margin is it the largest of the other terrestrial planets. True, it does possess a moderately thick atmosphere that overlies a thin patchy layer of water and it does have a noble satellite, about one-fourth its diameter. These qualifications of the Earth, however, are hardly sufficient to bolster our cosmic egotism. But, small as is the Earth astronomically, it is our best-known planet and therefore deserves and has received careful study.

Before the days of artificial satellites the dark part of the new Moon served as our best approximation to a mirror in space for studying the Earth (see Fig. 40). Near the new phase, when the Moon lies almost in a line with the Sun, the light reflected from the Earth illuminates the otherwise unlighted black hemisphere. Measures of the earthshine on the Moon indicate that the Earth is a good reflector of light, as are the other planets with atmospheres. Because of the variable cloud cover, however, the fraction of sunlight

Fig. 40. Earthshine on the Moon in an early phase. (Photograph by the Yerkes Observatory.)

Fig. 41. Southwestern North America as photographed by a U.S. Navy Aerobee
rocket from an altitude of 100 miles. (Official U.S. Navy photograph.)

reflected from the earth, its *albedo,* varies over a range of about
a factor of 2, averaging 0.35.

The remarkable photographs taken by rockets (Fig. 41), by the
U.S. Weather Satellite TIROS, and by several other U.S. spacecraft show the Earth from outside (see Fig. 42). As expected, the
clouds form the most conspicuous features at any time (Fig. 43).
The great extent of some of the cloud banks, however, proved a
surprise to the meteorologists; great waves of weather on the Earth
stand out in the cloud pictures. Meteorological satellites, with their
wealth of data on the worldwide state of the atmosphere, are
changing the *art* of weather forecasting into a *science.*

A hypothetical visitor at some distance in space, if equipped with a
good telescope, could eventually distinguish the continents from the
oceans and identify the polar caps. During the winter season in our
Northern Hemisphere he would see the north polar cap covering
an enormous area, some 50° from the Pole, while in summer the
area would shrink to only a few degrees of latitude. The lower

Fig. 42. The Earth as seen from space: a view of Central Africa from spacecraft Mercury-Atlas 4, showing Lake Rudolf in Kenya. (Courtesy of the National Aeronautics and Space Administration.)

border would always be very irregular, particularly where it is broken by the oceans. The south polar cap would change very much less because of the scarcity of land. The seasonal changes from green to brown to black and white in the temperate zones could probably be recognized and their causes explained by a clever observer.

One peculiarity that we cannot observe on any other planet could be seen by our hypothetical astronomer outside the Earth. He would be able to observe the *direct reflection* of the Sun from our oceans, when the Earth was properly turned. The phenomenon might be a great surprise for a Martian astronomer who had never encountered large bodies of water. He might very well attribute the bright reflection area to a smooth crystalline surface on the Earth, as the early astronomers visualized the Moon to be a perfect crystal sphere.

One observation about the planet Earth, as recorded by an outside astronomer in his book of facts, would be that the axis of rotation is not perpendicular to the ecliptic, the plane of revolution about the Sun. By long and careful measures he would conclude that

Fig. 43. The Earth as seen at an altitude of 23,000 miles by the U.S. applications Satellite 1 on December 9, 1966. Baja California shows in the upper right center and the solar reflection as a haze to left and below center. (Courtesy U.S. National Aeronautics and Space Administration.)

the equator is tipped 23.5° from the plane of the ecliptic. This *obliquity of the ecliptic* might enable him to account for the seasonal changes of color in the temperate zones and the variation in the sizes of the polar caps.

He would conclude that the direction of the axis remains fixed in space as the Earth moves in its orbit about the Sun, as in Fig. 44. When the North Pole was tipped toward the Sun (*a*), that hemisphere would be more illuminated by the Sun's rays. The Pole would be lighted continuously and the length of the day would be greater everywhere north of the equator. In addition, the light would fall on the surface at an angle more nearly perpendicular so that a given area would receive more heat and light during the daylight (Fig. 45), and would have more hours of sunlight.

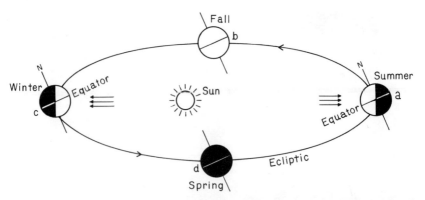

Fig. 44. The position of the Earth in its orbit during the four seasons. Winter and summer are both moderated in the Northern Hemisphere and accentuated in the Southern Hemisphere by the eccentric position of the Sun. The seasons, labeled for the North, are reversed for the South.

A quarter of the period (3 months) later, the Earth would be in position *b*, Fig. 44, and everywhere the hours of daylight would equal those of darkness. During the next half period (6 months) the Southern Hemisphere would gain in heat while the Northern Hemisphere would lose, the North Pole being completely dark during that time. If our outside astronomer possessed any ingenuity at all (and of course he must, being an astronomer!), he would be able to explain completely the observed changes in color in the two hemispheres, and also the peculiarities in the changes of the polar caps. He might express himself as follows: "Clearly on the planet Earth there must be complicated chemical or physical reactions that are directly activated by solar heat. Some regions, the dark-blue areas that constitute most of the planet's surface, are affected only by very great variations in heat while other regions, those that turn green as the temperature rises, are affected by much smaller

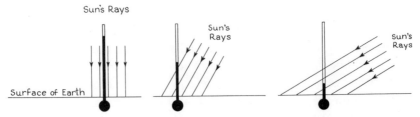

Fig. 45. The Sun heats a surface more effectively when the rays fall vertically, rather than obliquely.

changes. The permanent polar caps are probably similar to those regions that turn green with increase of temperature, but are never heated sufficiently for the reaction to occur."

Our learned friend from outside might continue, "We must conclude, therefore, that the more stable, dark-blue areas are very good conductors of heat as compared with those unstable areas which are so affected by slight changes . . ." Since the present writer is not too certain as to the remainder of the weighty conclusions, we may well let the matter drop at this point. The observations from outside would be of much more interest than the conclusions.

The time of highest temperature would be observed not to occur at the time of greatest length of day and of maximum sunlight on the surface. In the North Temperate Zone the maximum sunlight falls near June 21 (point *a* in Fig. 44), but midsummer, the time of highest temperature, comes actually late in July or near the first of August (in the Northern Hemisphere). The other seasons are correspondingly late. The seasons *lag* because the surface of the Earth (only the upper few feet and the atmosphere) becomes warmer as the amount of heat received from the Sun increases. The temperature continues to rise as long as the heat is strong, even though it is beginning to wane, until the rate of gain equals the rate of loss. Similarly, the coldest winter weather comes a month or more after December 21, the shortest day of the year.

It is interesting to note that the Earth is at perihelion, nearest to the Sun, during midwinter in the Northern Hemisphere and at aphelion, farthest away, during midsummer. The effect is to moderate the seasons slightly in the Northern Hemisphere, but to increase the range in temperature slightly in the Southern Hemisphere. In fact, however, the seasonal temperature changes in the Southern Hemisphere are smaller than in the northern, for a different reason. The oceans tend to control and moderate temperature changes in the atmosphere; the ratio of water area to land area is much greater in the Southern than in the Northern Hemisphere.

The Earth is generally called a sphere but actually is not a perfect one. Careful measures show that the diameter at the equator is about 27 miles (one part in 298.3) greater than the diameter at the poles; the equatorial cross section is bulged outward. This deformation does not arise by chance; the Earth was not cast in that shape to remain so forever. The internal gravitation is great enough to draw the material of the Earth into an almost perfect sphere, were there no rotation. The rotation in 24 hours, however, produces a centrif-

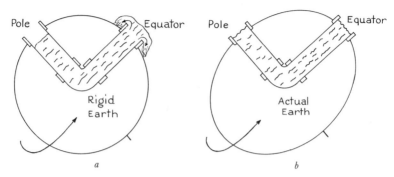

Fig. 46. (*a*) On a rigid spherical Earth in rotation, water would spill out of a pipe running from the pole to the equator. Although no such pipe exists, the oceans, correspondingly, would flow to the equator. (*b*) The actual Earth in rotation bulges at the equator, and hence the water level is uniform between the pole and the equator.

ugal force that enlarges the equatorial diameter at the expense of the polar diameter, to produce the observed equatorial bulge. If the underlying material of the Earth were too rigid to take the form prescribed by the rotation, the water in the oceans would flow to the equator to compensate the centrifugal force (Fig. 46*a*). Since the oceans are not particularly deeper there than elsewhere on the globe, we must conclude that the "solid" Earth adjusts itself to the force (Fig. 46*b*).

The presence of an equatorial bulge on the Earth, besides leading to such paradoxes as "The Mississippi River runs uphill," has one effect of great importance to the astronomer. This effect is called the *precession of the equinoxes,* observed in antiquity and explained by Newton. The term "precession" describes the fact that the Earth's axis does not remain fixed in direction over long intervals of time, but moves slowly around with a period of about 26,000 years. The angle between the equator and the ecliptic does not change essentially, although the axis twists around like the axis of a spinning top. The analogy (Fig. 47) is almost perfect, for the Earth really acts as a huge top.

The axis of the top is the polar axis of the Earth, the main body of the top is the Earth, and the rim of the top is the equatorial bulge. Because of the obliquity of the ecliptic, the bulge is always being attracted by the Moon, the Sun, and the planets, which try to turn the bulge, and therefore the equator, into the plane of the ecliptic. In the case of the top, the action of gravity is the reverse, tending to overturn the axis of rotation. In neither case does the overturning force succeed in causing the spinning body to tip over. Instead, the angle between the spin and the force remains the same but the axis precesses around as shown in Fig. 47. The peculiar property of a spinning body to resist a force applied to the axis is exemplified in the gyroscope, an instrument whose most important use is in gyrocompasses, in stabilizers, and in attitude-control units for ships, airplanes, rockets, and space vehicles. Gyroscopic control units are now amazingly precise, capable of being used as clocks as the Earth turns about them. They can hold orientation to the order of one part in a million.

The precession of the equinoxes ceases to be a purely academic problem when we look into the complications that it produces in calendar making. Back in Fig. 44 we see that the time of the seasons will depend on the direction of the Earth's axis. When the Earth and the Sun are on the line of the *equinoxes* (*b* or *d*), the line along which the planes of the equator and ecliptic intersect, the season will be either fall or spring. The precession of the equinoxes is a westward motion of the equinoxes as measured with respect to the stars, that is, clockwise when one looks down on the North Pole. If the year were defined as one revolution of the Earth about

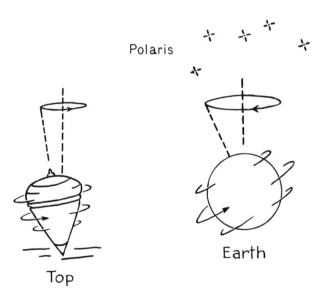

Fig. 47. A top and the Earth. Both precess because of forces that act to change the direction of their axes of rotation.

the Sun as measured with respect to the stars, the seasons would soon begin to get out of step with the months and within a few thousand years would be entirely changed. To avoid such a difficulty the calendar year, or *tropical* year, is measured from the time that the Sun is in the direction of the *vernal* (spring) equinox until it has returned there again. This tropical year keeps the calendar in step with the seasons but is shorter than the true, *sidereal,* year by about 20 minutes. The leap-year troubles of calendar making arise from the 5 hours 48 minutes 45.6 seconds that the tropical year exceeds 365 days in length.

Ancient astronomers noticed that the position of the stars in the sky at the same season shifted slowly with time, and also noticed the more obvious phenomenon that the North Pole of the Earth was shifting among the stars. Polaris, our present pole star, is only temporarily useful as such (see Fig. 180, star chart at end of book), though the motion of the pole in a lifetime is negligible. At the time when the Pyramids of Egypt were constructed, the pole star was α Draconis, some 25° from Polaris. The Southern Cross (Crux, a southern constellation) could then have been observed from most of the land that now makes up the United States. The pole moves in a small circle of radius 23.5° in the mythical "Year of the Gods." To these deities a man's life seems like a day. The period of precession is close to 70 × 365 years.

Since the Moon provides the major force on the Earth's equatorial bulge to produce the precession of the equinoxes, the bulge must, by Newton's third law of motion, affect the Moon's motion. It does; the major effect is to swing the plane of the Moon's orbit around (westward or backward) much in the manner of precession but in a period of only about 19 years. On nearby artificial Earth satellites the equatorial bulge has a profound effect, swinging the orbital planes around in weeks instead of years. Even though the computations of satellite motions are greatly complicated by these motions, they lead to some excellent new results. Not only is the oblateness of the Earth much better determined; the satellites show that the Earth is not gravitationally symmetric about the equator. It is slightly pear-shaped with the stem at the North Pole (Fig. 48). The distortion amounts to about 50 feet. Part of the pear shape arises from a pair of opposed gravitational "dimples" on Earth in the Northern Hemisphere, near India and off the west coast of North America. The equivalent depressions amount to about 230 feet and

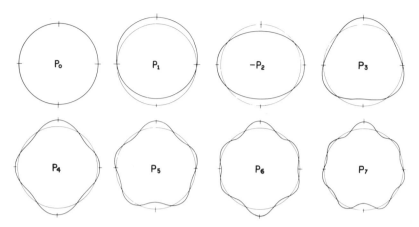

Fig. 48. The pear-shaped Earth is indicated by the diagram P_3, which represents only the third term in an infinite series giving a mathematical expression of the Earth's true shape. Artificial Earth satellites now give values for the first 21 terms in this series. The term $-P_2$ represents the Earth's oblateness. The North Pole is up.

are derived from density variations within the Earth's crust. Possibly they measure the effects of currents within the "solid" Earth (see Chapter 6).

To the astronomer the precession of the equinoxes presents many more serious difficulties than are involved in a mere change of the length of the year. He is forced to make his measures of celestial objects, natural or artificial, with respect to a reference system that does not remain fixed. His plight resembles that of an imaginary map maker who finds that all the continents and islands of the world are moving about. To state the latitude and longitude of a point on one of them would require a calculation involving the desired instant of time. Similarly in astronomy the starting point for practical measures is the vernal equinox, and the fundamental plane is the equator. Since these primary directions are moving because of precession, every published measure of a star, planet, or satellite must carry with it the date of the equinox and equator to which the measure is referred.

The motions of the Earth cause two additional complications in the problem of recording directions on the sky. One of these, *nutation,* is a small periodic irregularity in the procession, or the motion of the Earth's polar axis among the stars; it is caused by the Moon's peculiar motions and varying attraction on the equatorial bulge. The main effect of nutation is an oscillation in the motion of the

Earth's axis over a period of about 19 years. Only the most complicated mathematical theory enables the astronomer to calculate all of the small disturbances that finally add together in producing precession and nutation. Professor F. R. Moulton (1872–1952) expressed his admiration for the work in this field by saying, "No words can give an adequate conception of the intricacy or the beauty of the mathematical theory of nutation."

The second complication in observing the celestial bodies is the effect known as the *aberration* of light. Aberration was first observed and explained in about 1728 by the English Astronomer Royal, James Bradley (1693–1762). The story is one of the many fine examples of an exciting type of scientific research in which one phenomenon is sought and an unexpected one found. To prove that the Earth revolves about the Sun, Bradley attempted to observe the parallactic shift of the stars as the Earth moves about its orbit. He set his telescope very rigidly in a well-built (but disused) chimney in order to watch the daily passage of a certain star. If the Earth really revolved about the Sun, the position at each passage should have changed slightly during the year.

Painstaking observations failed to prove the motion of the Earth directly but did show a displacement out of phase with the one Bradley expected. He finally explained the new effect as due to the combined result of the motion of the Earth and the finite velocity of light.

There is a well-known story that the explanation came to Bradley while he was sailing on the Thames River. The vane on the masthead of the boat changed its direction as the boat changed its course although the wind remained steady from one direction. If we imagine the wind as being light from a star, the boat as our moving Earth, and the vane as a telescope pointing in the apparent direction of the star, we can see that the direction of the star will depend upon the motion of the Earth. In Fig. 49, where a telescope and an incoming light ray are shown, one can see that the motion of the telescope while the light is passing through will necessitate that the telescope be tilted forward in order to prevent the light ray from striking the sides of the tube.

A good example of aberration is the everyday experience of slanting tracks of raindrops down the side windows of a moving car. Drops that actually fall vertically may leave highly inclined tracks. When light falls upon the Earth from a star, the change in

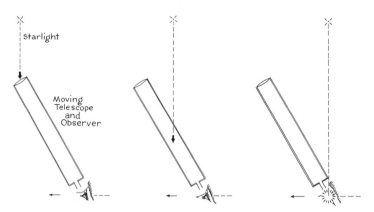

Fig. 49. Aberration of light. A moving observer must tip his telescope forward in order that the moving light ray may pass centrally through the tube.

direction by aberration is small, about 20 seconds of arc, but great enough that all observations of celestial bodies must be corrected for it. The velocity of the Earth is only 18.5 miles per second, as compared with the velocity of light, 186,282 miles per second. The ratio of these two velocities corresponds to the aberration angle (its tangent).

It is noteworthy that though it was Bradley also who discovered nutation, he never did attain his original goal of directly proving the motion of the Earth by the parallactic shifts of the stars. More than 100 years of telescopic improvements were necessary before this result was attained. Bradley's proof of the Earth's revolution about the Sun was nevertheless a good one, despite the fact that it was not the proof he had expected to find.

Not until 1838 was the effect of the Earth's revolution observed directly. Friedrich Wilhelm Bessel (1784–1846) was able to measure a slight shift in the position of the star 61 Cygni as it changed direction when seen from opposite parts of the Earth's orbit. The stellar parallax (Fig. 50), is 0.3 second of arc, that is, the astronomical unit would appear to subtend an angle of three-tenths of a second as seen from 61 Cygni. The nearest star now known is Proxima Centauri for which the parallax is 0″.76. The distance in miles is directly calculated as $(1/0.76) \times 206,265 \times 92,956,000$ or 2.51×10^{13} miles. Such a distance is too great to be visualized but is better expressed in terms of a light-year, the distance that light (moving at 186,282 miles per second) travels in a year. Light

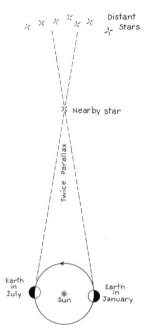

Fig. 50. The parallax of a star is the angle subtended by the radius of the Earth's orbit as seen from the star. The angle in the diagram represents twice the stellar parallax. The parallax can be directly measured for only the nearer stars because it becomes undetectable for the more distant ones.

comes to us from the Moon in 1.3 seconds, from the Sun in 8.5 minutes, but from Proxima Centauri it requires more than *4 years*. Even the bright stars, visible to the naked eye, are much more remote. The most distant star that can be photographed with the 200-inch telescope at Mount Palomar is hundreds of millions of light-years away, while the faintest group of stars, an *external galaxy,* is distant by thousands of millions of light-years. We can scarcely criticize the ancient astronomers for not anticipating these enormous distances.

The motion of the Earth is considered today not as the ponderous movement of a huge mass through space but as the natural movement of a small planet about an average star. In the depths of space the Earth counts for little more than a dust speck. To its inhabitants, however, the Earth is home, the Mother Planet, and is justifiably the most important member of the universe. In the next chapter we shall see how good a home it actually provides.

6

The Earth as an Abode
for Life

The Earth is taken for granted by the majority of its inhabitants. There is, of course, some grumbling about the bad weather, the poor crops, or the occasional catastrophes, but generally no critical analysis. Such an attitude was once justified by the fact that there was no alternative; having been born on the Earth we had no choice but to accept what hospitality it offered. With the conquest of space, however, we may now wish to consider whether it might be desirable to move to another planet. Let us, therefore, adopt a broader viewpoint and look at our Mother Planet with a critical eye. Let us consider the degree of security the Earth offers, the dangers lurking in space, and the unique conditions required for the maintenance of the fragile force we call *life*.

Let us first consider the dangers from without. For the existence of life as we know it, the immediate temperature must, some of the time, rise above the freezing point of water but must never exceed the boiling point. This restriction on the conditions of temperature is more limited than it seems at first glance, because the temperature scale begins at absolute zero, $-460°F$, and rises indefinitely. The highest temperatures directly observed on stars are some hundred thousands of degrees, while the interiors of the stars possess temperatures of many millions of degrees.

The Sun provides the Earth with the necessary heat to maintain its temperature within a suitable range, only 180 degrees out of millions, and does not raise the temperature too high. Evidence from the past indicates that the Sun has not *greatly* changed its output of heat for some hundreds of millions of years. The energy of the Sun arises not from any burning process but from nuclear fusion, the transformation of hydrogen into helium, in part by a complicated process involving carbon and nitrogen. The Sun's brightness should not change greatly for a *thousand million years* or more. However, a slight change of only a few percent in the Sun's heat would produce violent changes in the climate of the Earth.

Conceivably such changes in the Sun's radiation have been responsible for the great series of ice ages that have recurred every hundred million years or so. At the moment we appear to be within such a geological period, the ice having retreated only temporarily. The ice ages, however, represent no likely threat to life on the Earth, although they can make certain areas of the Earth uninhabitable for long intervals of time.

The Earth's atmosphere is a vital agent in maintaining a suitable temperature. It acts as a blanket to keep the noon temperature from rising too high and the night temperature from falling too low. Exactly as the glass in a greenhouse transmits the visual light of the Sun but prevents the passage outward of the heat or far-infrared light, to maintain a higher temperature than exists outside, so the atmosphere maintains a temperature balance near the surface of the Earth.

On the Moon, for example, where there is no atmosphere, the midday temperature surpasses the boiling point of water and the night temperature falls to about $-260°F$, much below the melting point of "dry ice." In space, outside the atmosphere of the Earth and well away from it, the temperature in the shade approximates the absolute zero, so that a heat-regulating atmosphere would be requisite to any active form of life in space.

The atmosphere, moreover, is a protecting roof from more than extremes of temperature. It is an invaluable shield from the meteors continually bombarding the Earth from interplanetary space (see

Fig. 51). These meteors meet the Earth at velocities up to 45 miles per second. A meteoric particle weighing only 1/30,000 ounce (1/1000 gram), moving at this speed, would strike with the same energy as a direct discharge from a 45-caliber pistol fired at point-blank range. Such a particle would be no larger than a fair-sized speck of dust, much smaller than an average grain of sand, yet dangerous to a person. Thousands of millions of such particles strike the Earth's atmosphere daily, constituting faint meteors that can be seen only with a telescope. The meteors visible to the naked eye are several times as large. In the atmosphere these bodies are quickly vaporized by friction with the air.

In space near the Earth's orbit the tiny meteoritic particles scatter sunlight to produce a glow seen near the Sun just before sunrise in the morning or just after sunset in the evening. Because the dust is concentrated toward the ecliptic, the glow is called the *zodiacal light* (Fig. 52).

Fig. 51. Meteors flash across the field of a fixed telescope. (Photograph by J. S. Astapovitsch.)

Fig. 52. The zodiacal light; the thin vertical line is instrumental. (Photograph by D. E. Blackwell and M. F. Ingham from Bolivia at an altitude of 17,100 feet.)

It is indeed fortunate that we are shielded from the meteors, but even so, some of the more massive ones are able to penetrate to the surface of the Earth and produce damage. The Great Barringer Meteor Crater in Arizona was formed by the explosion of such a huge body from space. This crater is nearly a mile in diameter and even now is nearly 600 feet deep despite infilling by erosion. The New Quebec Crater (Fig. 53) in Canada is much larger. Small meteorites have been found in abundance around the Barringer crater but no large ones have been discovered either by drilling operations or by radio-detection apparatus. The meteoritic body exploded at impact with a force far exceeding that of any known explosives. Only the "shrapnel" and the crater are left to tell us the story. The Great Siberian Meteor of 1908 detonated so violently that trees were laid flat up to 30 miles from the area of impact (Fig. 54).

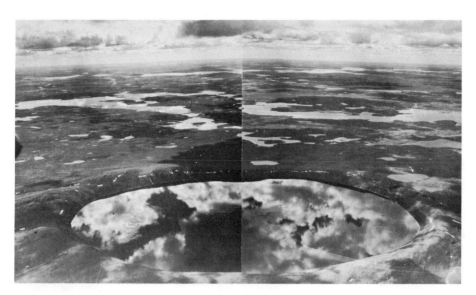

Fig. 53. New Quebec meteorite crater, nearly 3 miles in diameter. (Photograph by Royal Canadian Air Force; courtesy of the Dominion Observatory.)

Fig. 54. The Great Siberian Meteor devastated the forest over a distance of 20 miles from the point of fall. (Photograph by L. A. Kulik.)

Should another such meteor (or possibly comet) strike a large city, the damage would be comparable to that of a great nuclear bomb. Our only protection from such devastating meteors lies in their extreme rarity, but there is always the remote possibility that one of them might be encountered at any time (see Fig. 55). On the average a person is killed by a falling meteorite once in several hundred years.

Our atmosphere not only protects us from smaller meteors, it also guards us from death-dealing radiations in space. Light in the near ultraviolet causes sunburn but is generally important for health, although not a necessity. Ozone (three atoms of oxygen per molecule and producing the odor around electric discharges) is formed in the atmosphere by the Sun's light; it constitutes a shield from the rays of shorter wavelengths farther into the ultraviolet, where the rays begin to become dangerous to health. The oxygen, nitrogen, and other elements in the atmosphere cut out all of the far-ultraviolet rays below the limit of ozone. Such rays are sometimes used medically to kill bacteria in the air. Could they *all* reach the Earth it is doubtful that life in any form could exist.

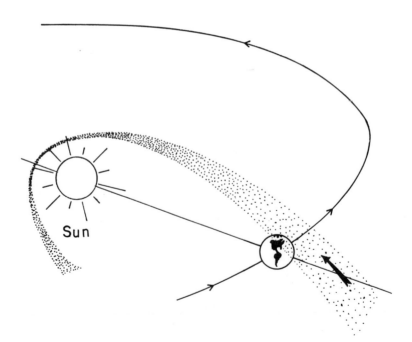

Fig. 55. A stream of cometary debris, moving in an elliptical orbit about the Sun, produces a meteor shower when the Earth passes through it.

Fig. 56. Aurora: a rayed band. (Photograph by V. P. Hessler at College, Alaska.)

Besides the far-ultraviolet rays, many particles that would be dangerous to life are stopped by the atmosphere. These particles, positive and negative in charge, emanate both from the Sun and from space in general. The streams of charged particles that cause the northern lights, or *aurora borealis* (Fig. 56), come from the Sun, while the *cosmic rays* come partly from the Sun and partly from unknown sources in distant space. There are other kinds of rays and particles in space that our atmosphere prevents from reaching the Earth. In the laboratory the cyclotron can now generate exceedingly dangerous streams of highly energetic particles equivalent to cosmic rays. Cosmic rays striking the earth and nuclear disintegrations in space produce high-energy radiations such as x-rays and gamma rays, which can be equally lethal.

It is clear that life as we know it requires a very specific set of circumstances for its continued existence on the Earth. The dangers from space specify that the planet must be within a rather definite distance of a star whose light is quite stable over long intervals of time, and the planet must possess an atmosphere capable of regulating the temperature and of screening out dangerous rays and particles.

No mention has been made of the exact composition required for the atmosphere. Until many more experiments have been made, the limits through which the composition may range without eliminating all possible forms of life are quite uncertain, but probably they are very broad. Oxygen, nitrogen, and carbon dioxide are probably essential components, while water must be available (see Table 1 for the composition of air). Surface water is not indispensable for certain desert plants but water in some form is necessary for all life as we know it. Volcanic gases and chemical combinations with surface rocks probably first determined the composition of the atmosphere. More oxygen and carbon, for example, are now combined in the rocks of the Earth's surface than exist in the air. When life became prevalent, the chemical reactions of the life processes affected the atmospheric composition; plants on land and in the sea as well as silicate rocks now compete for atmospheric carbon dioxide, but the balance is kept fairly stable by the decay of plant life and by the erosion of limestone rocks.

Since World War II high-altitude rockets have been used extensively to sound the upper atmosphere and nearby space (see Fig. 57). Instruments aboard radio their measures down to earth. They show

TABLE 1.   *Composition of air.*

Element	Percent by volume
Nitrogen	78.09
Oxygen	20.95
Argon	0.93
Carbon dioxide	.03
Neon	.0018
Helium	.0005
Krypton	.0001
Hydrogen	.00005
Water vapor	.2–4
Traces of other gases and dust	

Fig. 57. This Aerobee sounding rocket can carry 100 pounds of measuring instruments well above an altitude of 100 miles. (Official U.S. Navy photograph.)

that the atmosphere has practically a uniform composition to a height of about 60 miles. Winds of 100 miles per hour and more are observed in the meteor trains (see Fig. 58) and in the strange *noctilucent clouds* seen at the edge of the Arctic circle at altitudes even above 50 miles. Such strong winds mix the air sufficiently to prevent the light gases, such as hydrogen, from diffusing appreciably until above an altitude of about 60 miles. There are now several methods of measuring the density and temperature of the air at great altitudes. The methods are complicated, depending upon speeds of sound waves from explosions, upon the resistance of the air to meteors, upon the reflections of radio waves, upon rocket-borne laboratories, and, above 100 miles, upon the rate at which the resistance of the air brings down artificial satellites.

At low altitudes, up to about 20 miles, the temperature is measured by sending up small balloons with light meteorological equipment. These balloons carry tiny radio transmitters that send down messages of temperature as well as pressure and other characteristics, while their heights are being observed with telescopes or radars. The best estimates of temperatures at various levels are shown in Fig. 59, together with some of the phenomena that appear at these levels.

Fig. 58. An enduring meteor train, photographed at 10-second intervals, is rapidly distorted by high-altitude winds. (Photograph by the Harvard Observatory.)

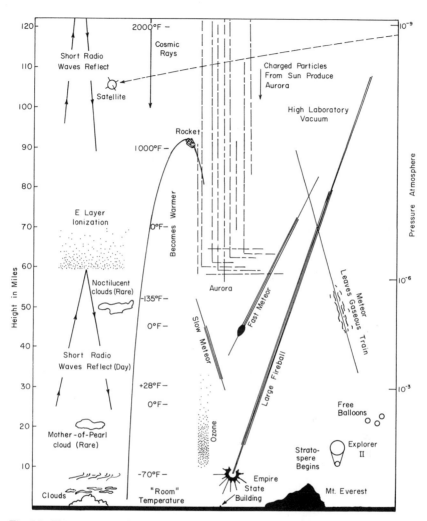

Fig. 59. Phenomena in the upper atmosphere.

The air temperature drops with altitude for a few miles in the range of most of our major clouds, where there is good vertical circulation and air cools as it expands on rising. With increasing altitude solar radiation causes a small percentage of the oxygen to change from breathable O_2 to the extremely poisonous O_3, or ozone. Ozone's absorption of ultraviolet sunlight heats the atmosphere and counteracts the falling temperature. Hence a minimum temperature occurs in the stratosphere, some 10 miles up, and a maximum some 30 miles up where nearly ground temperatures are recovered. Another very cold minimum occurs near 50 miles, above which the temperature begins to rise again. Above 200 miles a trace of atmosphere is left at a temperature of some 3000°F, heated by solar ultraviolet light and high-energy particles.

Above some 600 miles we encounter the famous Van Allen belts (Figs. 60 and 61), named for their discoverer, James A. Van Allen, who detected them by measures from early U.S. satellites. They may be classed as parts of the Earth's atmosphere, consisting of very energetic ionized nuclei of atoms, mostly hydrogen, and electrons trapped by the Earth's magnetic field. The Van Allen belts are very dangerous to living organisms not protected by shielding equivalent to about a half inch of lead, which some of the particles can penetrate. Some of the Van Allen particles are produced by cosmic-ray collisions in the atmosphere and the major portion by violent sprays of particles from the Sun in solar flares. They can also be introduced by nuclear explosions at great heights. From extreme

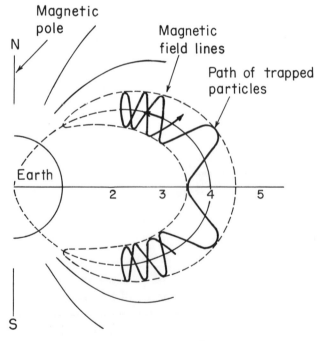

Fig. 60. Charged particles are trapped in the Earth's magnetic field because, as shown above, they are forced to turn as they move across the field lines. Numbers on the horizontal line indicate distances in Earth radii.

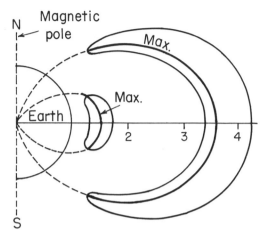

Fig. 61. The Van Allen belts of charged particles trapped in the Earth's magnetic field form two major rings around the Earth's magnetic poles.

parts of the belts the particles leak out fairly rapidly, in days or hours, while in the central part individual particles may persist for much longer times.

The air density decreases with height, reaching about one-millionth the surface value near a height of 60 miles. Half of the air is contained in the first 3.5 miles above the surface, half of the remainder above 7 miles, and so on. At the elevations where meteors are seen, radio signals are reflected, and the aurora borealis appears, the density of air is no greater than in the vacuum of a thermos bottle. A much thinner atmosphere than ours would still be a good shield from external dangers, but might not regulate the surface temperature so efficiently.

The interior of the Earth cannot, of course, be studied as easily as the atmosphere but there are methods of learning a great deal about it. Deep wells have penetrated only about 5 miles, a negligible fraction of the distance to the center. The astronomical effects from the equatorial bulge yield some information. Geological data are highly important in describing effects in the surface layers. Variations in the attraction of gravity from place to place and the strengths and directions of Earth magnetism add extremely valuable data concerning somewhat lower levels, but the most exact information about all of the layers deeper than a few miles proceeds from the manner in which earthquake (or seismic) waves travel through the Earth.

The many methods used to study the Earth are extremely ingenious and interesting, but even to describe briefly the more important ones would require no less than another book. A short outline of the results, however, will impart some impression of the Earth's construction and will demonstrate the precarious nature of the existence we enjoy on the skin of this planet.

Astronomical results give the average density (5.5 times the density of water) of the Earth and its shape at the surface; the attraction of the equatorial bulge and the known densities of rocks furnish a consistent idea of the densities of the Earth near the surface—about 2.8 times the density of water, or half the mean density.

The deep wells show that the temperature generally increases with depth at an average rate of about 1 Fahrenheit degree in 50 feet, although the rate varies tremendously from place to place. If this temperature increase were to continue to the center of the Earth, the temperature would reach the very high value of 400,000°F. That this temperature must be too great is shown by geophysical investigations, which place the central temperatures as low as several thousand degrees. The low heat conductivity of the surface rocks makes the change of temperature with depth very rapid near the surface, while in the deep layers heat is better conducted and the temperature changes more slowly. Radioactive elements, such as uranium and thorium, which seem to be concentrated largely in the Earth's crust, add to the heating of the outer levels. The residues of the radioactivity, such as helium and certain leads, provide a measure of the age of the rocks in which they are found and of the Earth itself—some 4.7 billion years.

Geophysical evidence indicates that the Earth was once heated, probably mostly by radioactivity, so that much or all of the interior was melted. As the short-lived radioactive atoms disintegrated, the entire planet cooled and most of it solidified. In the earlier stages, more than 4 billion years ago, the rate of heat loss must have been very much greater than at present because swift convection currents in the interior would have carried the heat up rapidly. After the formation of a solid crust the process became relatively very slow, volcanoes and lava flows carrying at most a few percent of the whole.

Without the information gained from the study of earthquake waves the structure of the deep interior of the Earth would be mostly conjecture. When an earthquake is produced, usually by a sudden slipping of the crustal rocks along a fault sometimes hundreds of miles below the surface, two main types of wave are sent out in all directions through the Earth. The *P*-wave, *primary* or *pressure* (or condensation) wave, is like a sound wave. The vibration is carried forward by a condensation of higher pressure and density, as in Fig. 62 (*upper*). The second type of vibration is the *S*-wave, *secondary* or *shear* wave, in which the motion is perpendicular to the direction of travel, as for a light wave or a wave on the surface of water (Fig. 62, *lower*). The *P*-wave, always traveling faster (approximately 1.8 times) than the *S*-wave, makes an earlier record of an earthquake on the seismograph at the recording station (Fig. 63). A few miles below the Earth's surface the *P*-wave travels at about 5 miles per second and the *S*-wave a little less than 3 miles per second. The speeds of both waves increase at greater depths, where the densities and pressures are greater. The destructive energy of an earthquake is carried by slower surface waves (*L*-waves), more complex in nature than the *P*- and *S*-waves.

The most remarkable result obtained from a study of the records made by these waves after they pass through various parts of the

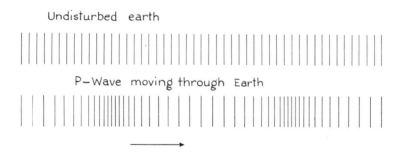

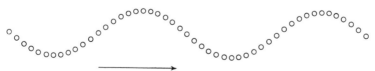

Fig. 62. Earthquake waves. The *P*-waves move by condensation and rarefaction, while the *S*-waves are transverse vibrations.

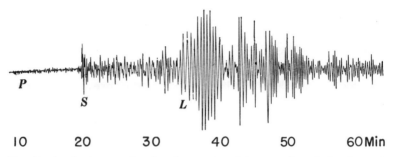

Fig. 63. A seismogram, showing the nature of the record made by the various earthquake waves. The *P*- and *S*-waves precede the stronger surface *L*-waves. (Courtesy of L. Don Leet.)

Earth is that the *S*-waves do not penetrate a central core (the Dahm core), which extends slightly more than half way out from the center (Fig. 64). Since *S*-waves are damped out in liquids, it is thought by many geophysicists that the Earth has a *liquid* core. It is possible, however, that the core is really solid and that a liquid or peculiar layer at its surface stops the shear waves. Whether actually liquid or not, the core differs considerably in structure from the levels above it, having a density about twice the average. Some 800 miles from the center an inner core of very high density has recently been discovered. It may be solid.

The pressures near the center of the Earth are tremendous, about 50 million pounds per square inch. It is difficult, therefore, to predict at what temperature any given material would melt or how much it would be compressed. The compression of a solid or liquid is certainly considerable near the center and the melting temperature will certainly be high. Since the Earth is magnetic and since iron and nickel-iron are so prevalent in meteorites, which are our only samples of material from other planetary bodies, most investigators believe that the core of the Earth is made largely of iron or nickel-iron. The high pressure compresses the iron from a density of 7.7, which it has on the Earth's surface, to about 10 to 12 at the center. There the *P*-waves travel about 7 or 8 miles per second.

Just outside the inner cores in the intermediate shell as seen in Fig. 64, the densities are about the average for the entire Earth. Some iron-stone mixture under pressure may constitute this layer or perhaps the pressure compresses ordinary rocky material to this density. The outer mantle is somewhat denser (4.3 times water) than the heavier rocks, but may be largely composed of them. The

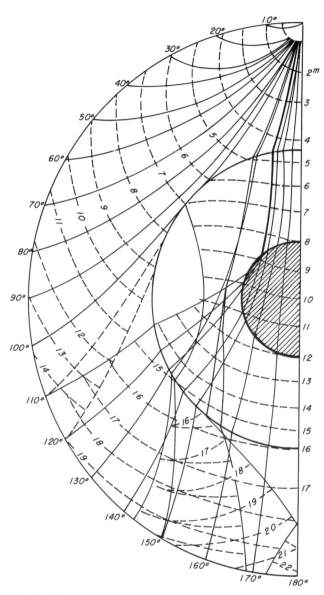

Fig. 64. Paths of earthquake waves in the Earth. The changing density with depth causes the paths to curve. At the discontinuities of the core boundaries the *P*-waves are bent abruptly. The *S*-waves do not penetrate the core and are shadowed from stations opposite the earthquake. The numbers represent minutes of travel time. Note the newly discovered solid (?) inner core.

crust consists mostly of granites and other igneous rocks, the sedimentary rocks appearing only in the upper mile or so, on the average.

The outer crust of the Earth varies in thickness from place to place and from investigator to investigator, but appears to be some 20–25 miles thick under the continents and as thin as 3–4 miles at some spots under the oceans. Thus there is hope of actually boring through, in the proposed *Mohole* experiment. Geologically the crust appears to float on a deformable but exceedingly viscous layer perhaps 100 or 200 miles in depth. The formation of mountains and the extensive distortions of the crust show clearly that the crust is subject to motions that could not be possible unless there were an underlying layer of pseudo-liquid material that can yield. According to the principle of *isostasy* the total mass under any given area is constant. Lighter surface materials, such as those that form mountains, are lifted. Glaciers depress the surface, which again rises slowly after the ice melts. For quick-acting forces, such as earthquake waves, the material is very rigid; to forces acting over long periods of time it gives way. Glass is such a material. "Silly Putty," a silicon compound that bounces on impact but kneads like putty and flows like water in a few hours, simulates the properties of the solid earth, on a very short time scale. Volcanoes show, of course, that some liquid material must exist just below the crust, but all of the deformable layer need not be liquid in the ordinary sense.

In recent years extensive mapping, sampling, and temperature measuring of the ocean bottoms have been made, particularly by the Woods Hole Oceanographic Institute in Massachusetts and the Scripps Institute at La Jolla, California. Down the whole length of the Atlantic Ocean occurs a high ridge with a deep rift near its peak (Fig. 65). Such a system of ridges, as high as a great mountain chain and usually with a rift, occurs in most oceans. Generally the temperature gradient in the ocean bottom is greater near the rifts, indicating a greater heat loss there and a thinner crust. Horizontal faults tens of miles in length often break the continuity of the rifts.

Some geoscientists speculate that these great worldwide systems may represent the surface effects of huge heat-convection cells extending deep into the mantle of the earth. Conceivably they lie dormant for many millions of years while erosion wears down the continents and the climate becomes more uniform on a world-wide basis. The heat accumulating from radioactivity in the rocks reactivates the convection cells, which carry the internal heat to the sur-

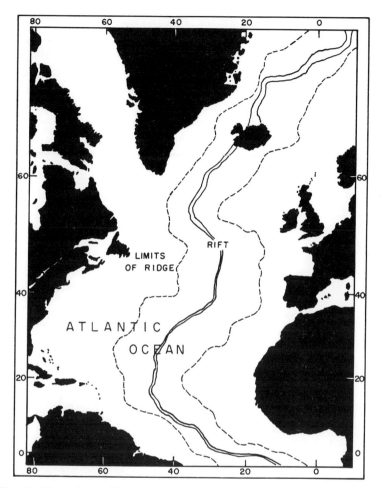

Fig. 65. The Atlantic ridge, with the floor of the rift valley 9,000 to 15,000 feet below sea level flanked by ridge peaks submerged only 4,000 to 7,000 feet. Deep earthquakes occur systematically along the rift. The numbers are degrees of latitude and longitude. (After Bruce C. Heezen.)

face and build new mountain chains. The mountains and irregular continental areas then upset the worldwide air circulation and climate. Perhaps the great geological ages recur in this fashion roughly every 100 million years. Exactly how the ice ages occur is not clear but they are certainly associated with mountains and mountain building.

Two pieces of geological evidence—climatic changes over the last 500 million years and records of the ancient magnetic field in old sediments—indicate that the crust of the Earth has slipped or slid around the interior like the shell of an egg. This motion is reminiscent of the classical method of distinguishing between fresh and hard-boiled eggs without breaking the shells. A spinning fresh egg that is stopped and suddenly released will start spinning again. The North Pole may once have been in the Pacific Ocean. The Earth's pole does have a wobbling motion of a few feet as measured by changes in the latitudes at various stations and possibly may be moving systematically (Fig. 66). That the continents have moved with respect to each other no longer remains uncertain. Figure 67 illustrates the manner in which the various continents *may* have

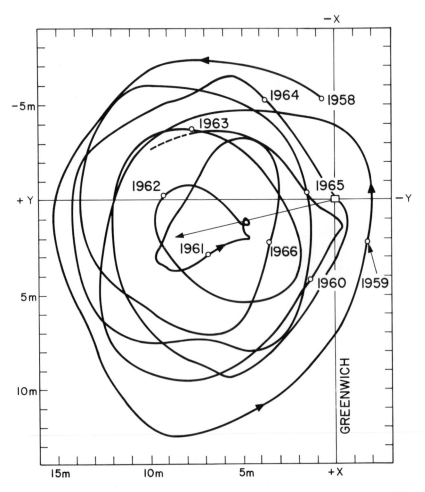

Fig. 66. Motion of the North Pole over the Earth's surface from 1958 to 1966. The rectangle measures the mean starting point in 1903, and the arrow indicates the average direction of motion, some 4 inches per year. (Courtesy Smithsonian Astrophysical Observatory.)

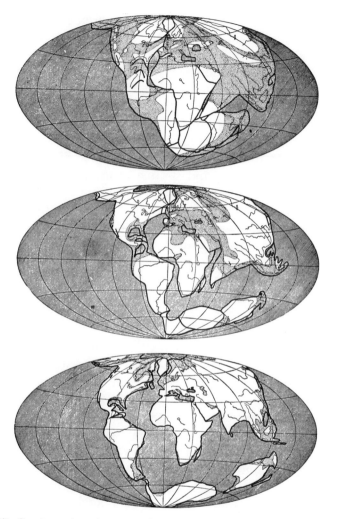

Fig. 67. Continental motions according to Wegener's theory. (From *Lehrbuch der Physik;* courtesy of Friedrich Vieweg und Sohn.)

developed from a single continent. The fit of the "jigsaw-puzzle" pieces is supported by new evidence in addition to biological evidence. The oceanographic studies, mentioned above, show that sediments are thinner and newer near the mid-oceanic ridges. Certain old rock formations in Africa and South America, as well as in Greenland and the British Isles, match up well with the boundary shapes shown in Fig. 67. The ocean bottoms near the ridges are clearly spreading apart—very slowly, a few inches per year. Thus the continents may well have broken away from a common nucleus eons ago and have spread apart and grown in the meantime. Lasers acting as radars on reflecting satellites, space probes, and the Moon should soon measure intercontinental distances to accuracies of feet and give us the answers sooner than we might expect. Observations of artificial satellites have already reduced these errors in distance from a hundred yards to ten yards!

The wobbling of the Earth's pole where it "pierces" the Earth's surface is a small effect, but of interest as it is well explained by theory. The principal motion has a period of about 430 days, while there is a smaller motion in a year (see Fig. 66). Seasonal changes cause a melting and shifting of the ice in the polar caps which would produce a small yearly effect, but the 430-day period requires more explanation. If we again consider the Earth as a top, it is like one that was set spinning badly, not exactly about the symmetrical axis perpendicular to the plane of the equatorial bulge. Major earthquakes may also tend to displace the pole slightly and to change the nature of the polar wobbling. If the Earth were absolutely rigid, the pole would oscillate in about 10 months, but since the Earth is only twice as rigid an steel, the period is 430 days, as observed.

The fragile crust of the Earth, floating on the heated and deformable rocks underneath, is not the stable and permanent layer that it appears from everyday experience. Not only is it clearly drifting about the main body of the Earth, but it is certainly cracking and buckling through geologic ages. Many regions, having become completely covered with ice in the glacial ages, sank under the load. When the ice melted, they rose again. Most of the present land areas have been under water for long periods of time, and some of the sea bottoms have been dry land. The magnetic field keeps wandering around and occasionally changing sign while the continents are shifting and changing shape.

Mountain ranges rise and are worn away by the rains. The crust is in almost continuous vibration produced by earthquakes that originate from shifts and readjustments in the various strata beneath, sometimes several hundred miles deep. Volcanoes may become active at any time and occasionally produce catastrophic results, such as the violent explosion of Krakatau. Furthermore, great meteorites occasionally fall and can devastate huge areas of the Earth.

When we contemplate all of these dangers to life both from within and without the Earth, we must indeed marvel that we still exist—but, of course, if we did not . . .

7

The Moon's Influence
on the Earth

The era is well past when mystical powers of the Moon were supposed to influence our everyday life on the Earth. No longer do thinking people attempt to credit the Moon with their successes or blame it for their failures. The Moon does, however, influence the Earth directly in many ways—all subject to simple laws of physics and dynamics.

The Moon is so large and so close to us that it reflects sufficient sunlight at its full phase to light up the night satisfactorily for many practical purposes of life. It is massive enough to distort the shape of the Earth and to produce tides in lakes and oceans. It provides the main force that moves the poles of the Earth in the precession of the equinoxes. Its distortion of the Earth's shape produces friction that slowly lengthens the hours of the day. Its shadow on the Earth at occasional places and times obscures the light of the Sun to produce solar eclipses. In such ways our nearest neighbor in space makes its presence known. To see how these effects are brought about, let us begin by investigating the motions and superficial appearance of the Moon.

The Moon's period of revolution about the Earth is approximately represented by the calendar month. Were fractional months feasible in a calendar, there should be 12.37 . . . months per year, because their average length is 29 days 12 hours 44 minutes 2.8 seconds. This period, technically the *synodic* month, is the space of time in which the Moon passes through its sequence of phases from *new* to *first quarter,* to *full,* to *third quarter,* to *new* again (Fig. 68), and makes a complete revolution about the Earth with respect to the Sun. Since the Earth has moved forward about 30° in its orbit during this time, the true or *sidereal* month, measured with respect to the stars, is a little more than 2 days shorter than the synodic month. On the average, a sidereal month has a length of 27 days 7 hours 43 minutes 11.5 seconds.

The reason for the difference in lengths of the two months can be seen from Fig. 69. Starting from a new moon, *A,* when Sun, Moon, and Earth are in line, we see that the Moon returns to the same direction with respect to stars *before* it reaches the *new* phase again, because the Earth has completed part of its revolution about the Sun in the meantime.

The most curious fact about the Moon's motion is that the Moon

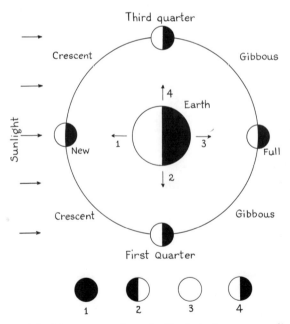

Fig. 68. Phases of the Moon as seen from the Earth during one synodic month.

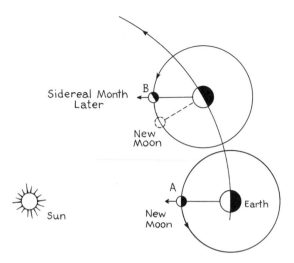

Fig. 69. A sidereal month is shorter than a synodic month. The Moon is not yet new again at position *B*, although it has completed a revolution about the Earth with respect to the stars.

rotates on its axis at the same average rate that it revolves about the Earth. Thus we always see the *same* hemisphere of the Moon's surface and *never* can see the other hemisphere. To demonstrate this motion, hold a ball or globe rigidly at arm's length and slowly turn around. As your body makes one revolution so does the ball, but you see only one side unless you turn the ball in your hands.

In actuality, we have an opportunity to peek around the rim of the Moon to a certain extent, mainly because the Moon revolves in an elliptic orbit. As a consequence of the ellipticity, the rate of revolution is not uniform, although the rate of rotation is nearly so. In addition, the orbit is tipped some 5 degrees to the ecliptic, its pole moving westward around the pole of the ecliptic in about 19 years. Thus, we can also peek over the poles of the Moon by about 5 degrees. Our location on the Earth enables us to see a bit more of the Moon; moreover, it oscillates very slightly as it rotates, through 1 or 2 minutes of arc in about 3 years. These effects that enable us to see some of the Moon's "forbidden" hemisphere are called *librations*. When all of the librations are summed up it is possible for us to see 59 percent of the Moon's surface at one time or another, while 41 percent cannot be seen at any time. The conquest of space has now given us the secret of the Moon's other face (see the next chapter).

Since the Moon is the nearest celestial object, its distance is the most accurately known. The older method of measuring the distance is almost exactly that used for Eros, as described in Chapter 4. Radar now gives us a confirming value, accurate eventually to yards. The nearest that the Moon can approach the Earth's center is some 221,463 miles. An observer, of course, may move some 4000 miles closer than this because he is necessarily located on the surface of the Earth. When one sees the Moon overhead he is closer by about 4000 miles than at moonrise or moonset (Fig. 70). The greatest distance that the Moon can attain is 252,710 miles, while its mean distance is 238,856 miles.

The Earth's atmosphere has a surprising effect upon observations of the rising or setting Moon. Light rays are bent by the atmosphere to such an extent that the *entire Moon (or Sun) can be seen before it has risen and after it has set.* The *refraction* of the light coming from empty space into the atmosphere is just about 0.5 degree, the apparent diameter of the Moon. Thus, when the Moon's upper limb would be just out of sight were there no atmosphere, the entire Moon is apparently lifted into view (Fig. 71). At greater heights the refraction is less, and it decreases to zero overhead.

Everyone has noticed the strange phenomenon that the Moon appears to be larger when seen near the horizon than when seen overhead. Actually, when *measured,* the diameter is smaller when near the horizon, because of the small distance effect mentioned above and because refraction flattens the disk slightly. The standard

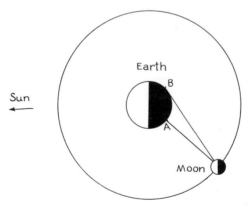

Fig. 70. The Moon's distance is less for observer *A*, who sees the Moon overhead, than for *B*, who sees it setting.

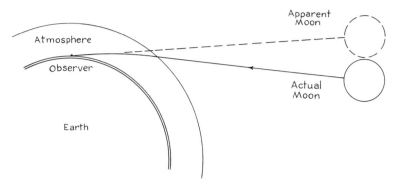

Fig. 71. Refraction in the Earth's atmosphere apparently raises the Moon (or Sun) above the horizon after it has geometrically set.

explanation has been that the Moon seems larger when seen in juxtaposition to distant objects on the horizon than when seen against the expanse of the sky. Since the effect is the same for an unbroken horizon at sea as for a land horizon, this explanation is not satisfactory. Psychologists show that the effect arises from a peculiar property of the brain and eye. The observer tends to visualize the Moon as more distant when near the horizon than when overhead. That is, he unconsciously places it at the distant horizon. Since the Moon remains unchanged in angular size it *seems* to be intrinsically larger near the horizon where it *seems* to be more distant (Fig. 72).

Important as the Moon may be as an object for stimulating the human mind, its greatest effect on the Earth results from its power to produce tides. This power is a direct consequence of the gravitational attraction of the nearby Moon for the Earth. The cause of the tides was early appreciated by Newton as a confirmation of his law of gravity. Since the attraction is inversely proportional to the square of the distance, the part of the Earth nearest to the Moon is attracted by a force nearly 7 percent greater than the part farthest away. The force at the center is, of course, the average value, which is exactly sufficient to hold the Moon in its orbit. The 7-percent differential in force acts on the body of the Earth as a distortion, tending to stretch the globe along the line joining it with the Moon (Fig. 73).

A most interesting feature of this tide-raising force is that the face of the Earth *away* from the Moon is distorted in almost exactly the

Fig. 72. The Moon illusion: the two black disks are of equal diameter. (After a diagram by I. Rock and L. Kaufman.)

same fashion as the face toward the Moon. One may understand this symmetric elongation by considering that the lunar hemisphere of the Earth is pulled away from the center, and that the center is pulled away from the opposite hemisphere. When the Earth is stretched along the line joining it to the Moon, the circumference perpendicular to this line is naturally compressed. The net tendency of the tide-raising force is to distort the Earth into a shape similar to that of a symmetric egg.

Now if the Earth were absolutely rigid, not yielding to the distorting forces that act on it, all of the tidal effects would occur in the oceans and surface waters. If the Earth were perfectly elastic with no rigidity, the ocean tides would be negligible, although the tidal bulge would still exist. The comparison of ocean tides with the predicted values is exceedingly difficult, however, because the measured tides at shore stations depend upon currents that are set up over the irregular ocean beds. Careful measures of the tides in long pipes show that only 70 percent of the theoretical effects actually occur. The main body of the Earth yields to the forces to the extent of the remaining 30 percent. From these measures it is deduced that the Earth as a whole is more rigid than steel. The data from observations of earthquakes and from the motion of the Earth's poles confirm this result. Outside the inner (possibly liquid) core the Earth has an average rigidity about twice that of steel.

A surprising additional result about the Earth was found from the tide experiments. *The Earth is an elastic ball.* Before the experiments, it was generally believed that the Earth was viscous, like thick molasses or glass; if it were distorted a small amount it would probably remain so or else *slowly* regain its original shape because of the small restoring forces. The experiments showed that the entire Earth yields *immediately* to the tide-raising forces, in so far as its rigidity will allow, and that it *immediately* returns to its original shape when they are removed. Thus the Earth is not only more rigid than steel; it is also more elastic.

Although the Moon is the most powerful body in raising tides on the Earth, the Sun also is an important contributor—to the extent of about 30 percent. The Sun produces tides in exactly the same manner as the Moon. When the two bodies are nearly in line, as at new or full moon, their tidal forces add together. When their directions are at right angles, as at first or third quarter, their tidal effects tend to cancel. The result is that at new or full moon there are *spring* tides, in which the high tide is very high and the low tide is very low. In between, at first or third quarter, there are *neap* tides, in which the range from high to low tide is reduced to less than half the value at spring tides.

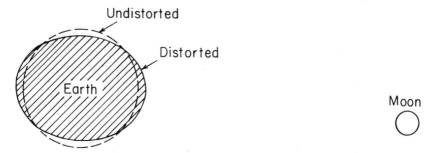

Fig. 73. The cause of tides. The Moon elongates the Earth along their line of centers.

Still another factor enters into the production of the tides. When the Moon is nearest to the Earth, at *perigee,* its tide-raising force is greater than when it is farthest away, at *apogee.* The range in the lunar part of the tide changes by about 30 percent because of this change in distance. The combination of the varying amount of the lunar tide and the varying summation of the lunar and solar tides causes large changes in the ranges of the ocean tides.

To see how the tides should occur in practice, we shall first take the ideal case when the actual high tide takes place exactly at the top of each of the tidal bulges shown in Fig. 73. In the ideal case there are two high tides each day as the Earth rotates and the observer passes the top of the two bulges. The high tides take place when the Moon is overhead (on the *meridian,* an imaginary north-south line on the sky) and when it is on the opposite side of the Earth, while low tides occur in between. Because of the revolution of the Moon about the Earth, the tides occur about 50 minutes later each day.

Twice a month, at new moon and full moon, there are *spring* tides, when the lunar and solar tides add together. In between the spring tides, at first and third quarter, there are the *neap* tides, when the solar tide subtracts from the lunar tide. Once a year, generally, the new moon occurs at a time when the Moon is near perigee, while about 6 months later the full moon occurs at this position. The resultant spring tides at these two times of the year are especially high, because of the increased size of the lunar tides. The dates of these maximum spring tides are progressively later by more than a month from one year to the next, because the direction of perigee is always moving forward around the Moon's orbit, with a period of nearly 9 years.

Although the prediction of the theoretical tides is somewhat complicated, the prediction of actual tides at a given station is even more difficult. The tides at shore stations generally have ranges of several feet, considerably above the average expectation by simple theory. This discrepancy arises from the fact that the observed tides are measured at the shallow edges of the oceans. As the Earth rotates, the tidal bulges of Fig. 73 become, in effect, tidal waves, which pile up on the sloping ocean beds near the shores, much as ocean swells may grow to high waves as they approach a gently sloping beach. In the Bay of Fundy, where this effect is further augmented by the occurrence of a funnel-shaped shore line, the tidal range is often 50 feet or more.

At any point on the shore the time required for the tidal wave to pile up to the maximum height depends entirely upon the contours of the ocean bed. In many places the high tide is consistently later than the maximum tide-raising force by several hours. This delay is known as the *establishment* of the port, and is used in predicting the tides. If one knows the phase of the Moon and the establishment for his position along the coast, he can usually estimate the times of high and low tides with an error not exceeding an hour. At new or full moon, the establishment represents the number of hours after noon or midnight (sundial time) that high tide should occur. At first or third quarter the same prediction gives the time of low tide. By adding a correction of 50 minutes for each day elapsed since the nearest preceding phase of the Moon, the times of the tides can be estimated at any intermediate date.

The obliquity of the ecliptic has a marked effect on the tides at stations away from the equator. Because the Earth's poles are tipped from the plane of the Moon's revolution, the two daily tides may differ greatly in range. By reference to Fig. 74, one can see that the tide at point A will be greater in range than the one at B, half a day later. At certain stations it often happens that only one high tide instead of two will be appreciable.

Along coastal regions the tides are naturally of great interest and are vital in the everyday affairs of seafaring people. For most of the other inhabitants of our globe, the tides are only an interesting phenomenon connected with the sea. In any case the tides are so taken for granted that the Moon's importance is apt to be forgotten. Practically no living person, on the other hand, can fail to be impressed by a second remarkable phenomenon caused by the

Fig. 74. Unequal daily tides. The tide at A will exceed that at B.

Moon—a total eclipse of the Sun. At rare intervals, because of a striking coincidence, the Moon is at just the proper position to blot out the Sun's light for a small area on the Earth. If the Moon were a bit nearer, solar eclipses would be commonplace, and if it were a bit farther away we could never scc a total eclipse.

In Fig. 75, the Moon's shadow on the Earth is shown. For an observer inside the dark cone (the *umbra*), no direct rays of the Sun are visible; only the *corona,* the outermost vaporous atmosphere of the Sun, and high prominences, can be seen. Outside the umbra, in the partially darkened shadow (the *penumbra*), a part of the Sun's disk is covered. As the Moon's shadow passes over the Earth's surface, the sunlight is slowly dimmed during a period of an hour or more (Fig. 76). As the light becomes weaker, with only a thin crescent of the Sun's disk exposed, a phenomenal coolness and silence prevails. The crescent is still so brilliant that it must be viewed through darkened glass. Just before totality the narrowing crescent breaks up into a series of beads, as the last rays from the Sun shine through the valleys of the precipitous lunar surface. These *Baily's beads* show brilliantly for a few seconds only (Fig. 77). By this time a glowing ring can be seen completely around the Moon, and if, as occasionally happens, one bead alone is bright, the effect is that of a luminescent diamond ring. Oddly enough, these beads really should be called Williams' beads, since Samuel Williams (1743–1817) observed and described them many years before Francis Baily (1774–1844) did.

Quickly the sky becomes dark as in the evening dusk. At this moment the corona seems to flare out in all directions about the Sun (Fig. 78). Long spikes of light, several Sun diameters in extent, with their bases in the glowing halo of light, point outward into

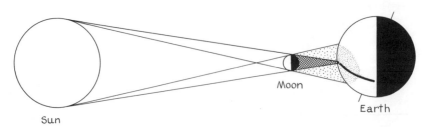

Fig. 75. Total solar eclipse. The dark shadow is the umbra of total eclipse and the shaded area is the penumbra of partial eclipse. A path of totality is shown. The relative dimensions are much exaggerated.

Fig. 76. Solar eclipse of February 1961. (Photographed by A. H. Catá of Geneva, Switzerland; courtesy of *Sky and Telescope*.)

the dark blue of the sky. A few bright stars and planets may be seen. The silence and coolness are awesome and the incandescent silver and blue of the corona are magnificent beyond the expression of words.

In a very short time the spectacle is shattered by the appearance of dazzling rays from the beads that appear on the west limb of the Sun. After totality, the beads appear much brighter and more striking than before totality, because the eye has been accommodated to the semidarkness. Soon the crescent of the Sun brightens the land-scape and the partial phase of the eclipse is slowly repeated in reverse order, until the Sun is completely unobscured.

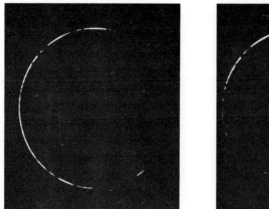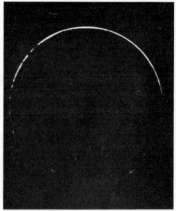

Fig. 77. Baily's beads. (Photograph by the Yerkes Observatory.)

A total eclipse can never continue for much over 7 minutes and usually lasts for a considerably shorter time, yet it is a sight that well repays an observer for his effort in traveling to the zone of totality. To the astronomer, it is an opportunity to observe the corona, which is an extended but exceedingly rarified and hot mantle of gas about the main body of the Sun. Also, the *prominences* of the Sun, great whirling or exploding clouds of hydrogen and calcium gases, can be seen (Fig. 79), though they and the corona can now be well observed without an eclipse. The astronomer has also an opportunity to photograph and measure the positions of stars near the Sun, where no measures can ordinarily be made because of the brilliancy of the sunlight scattered in the Earth's atmosphere. These measures have shown that the light from distant stars has been minutely deviated by the Sun's mass, in accordance with the predictions of Einstein's theory of relativity. This demonstration, coupled with the anomalous motion of Mercury's perihelion, constitute two of the three astronomical verifications of the relativity theory.

After many months spent in the construction and preparation of instruments, often after a long trip to a distant part of the Earth where the eclipse is to be seen, and after a strenuous and usually hurried final erection of the instruments at the chosen observing site, the astronomer is indeed happy if he is fortunate enough to have just a few minutes of clear sky at the critical moment. Although solar eclipses are fairly numerous, from two to five each year, such expeditions are necessary because the area of the Earth covered by the dark umbra of totality is very small; the width of the path is only a few tens of miles. In any given location a *total* eclipse can be seen, on the average, only once in 360 years. Sometimes, only the penumbra of the Moon's shadow strikes the Earth, and produces a partial eclipse, while in other eclipses the Moon is so far away that its disk does not completely cover the Sun. In the latter case the eclipse is *annular* or ringlike (Fig. 80). Annular eclipses are of some use in measuring accurately the profile of the Moon's limb.

Eclipses of the Moon by the Earth's shadow are less numerous than solar eclipses but each is observable over more than half of the Earth's surface (Fig. 81). Consequently, at a given position on the Earth, a lunar eclipse can be seen quite frequently. In some years

Fig. 78. The solar corona at the total solar eclipse of August 31, 1932, Maine. (Photograph by the Lick Observatory.)

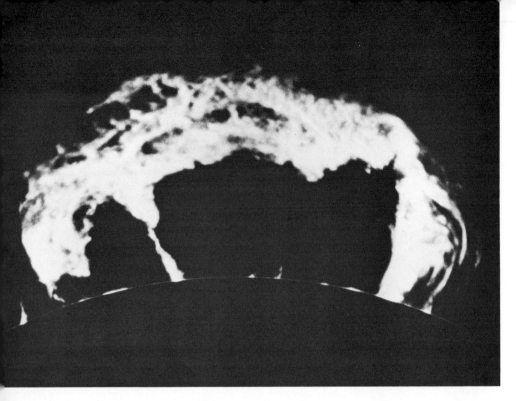

Fig. 79. Solar prominences are giant clouds of incandescent hydrogen, calcium, and other gases. They take many forms, which are controlled by magnetic fields on the Sun. (Photographs by the Sacramento Peak Observatory, Air Force Cambridge Research Laboratories.)

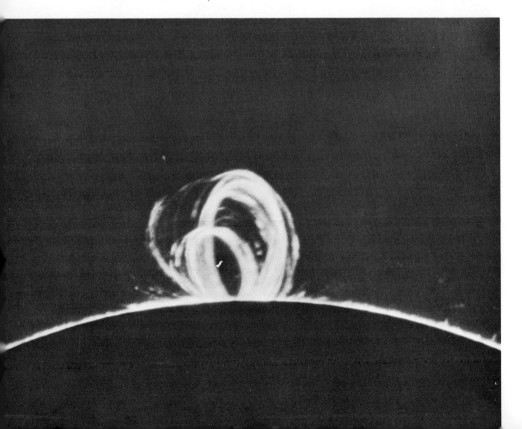

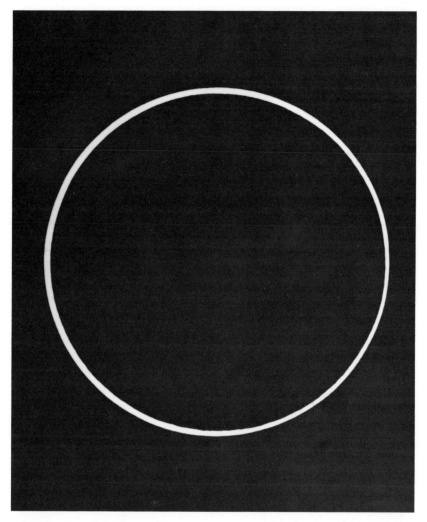

Fig. 80. Annular eclipse of the Sun, photographed at Senegal, July 30, 1962, by L. G. Stoddard and G. Moreton of the Lockheed Solar Observatory. Irregularities in the Moon's profile can be seen.

none occur, while the maximum number is three. The maximum number of eclipses in one calendar year is seven, five solar and two lunar or four solar and three lunar. A lunar eclipse, however, is not at all spectacular, and it has little value to the astronomer. When the Moon lies completely in the umbra of the Earth's shadow, it usually acquires a dull copper hue because some sunlight is refracted through the Earth's atmosphere to produce a sunset effect.

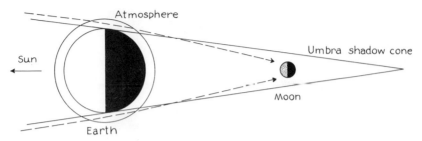

Fig. 81. The Moon is visible during a lunar eclipse because the Sun's rays are bent into the umbra by refraction in the Earth's atmosphere.

On rare occasions, the eclipsed Moon is very dark, because the Earth's atmosphere is clouded completely around the twilight zone; at other times part of the Moon is illuminated during totality (Fig. 82).

The ancient observations of solar eclipses, on the other hand, have been invaluable in showing that the Moon is tending to increase the length of the day by acting as a brake on the Earth's rate of rotation. The lack of accurate time-keeping devices in antiquity has been no handicap in this type of investigation, because the *place* where a total eclipse of the Sun could be observed is, in itself, a good measure of the time and of the position of the Moon when the eclipse occurred. The Earth must be turned at a certain angle and the Moon must be in a specified position for the Moon's shadow to fall on a given point of the Earth's surface. Calculations based on the records of ancient eclipses show that the day is *increasing in length* by nearly 0.001 second every century. This change must arise from some type of tidal friction.

The energy of the tides is thus changed to heat, at a rate of some 4 billion horsepower. In 1920 Sir Harold Jeffreys calculated that the friction of the moving water in the shallow areas of the Irish Sea accounted for 80 percent of the observed change in the length of the day. Recently, however, W. H. Munk and G. J. F. MacDonald question this conclusion and consider the problem unsolved. Recent seismic evidence and tidal experiments on land combine to suggest a solution that may eventually prove correct. The Earth's crust appears to divide into blocks a few miles in dimension. As tidal forces distort the crust the blocks react individually, slipping or "grinding" against each other. Thus the tidal friction may take place in the crust as well as in extended shallow waters.

Some remarkable research by John W. Wells of Princeton

Fig. 82. Total eclipse of the Moon. Part of the Moon is somewhat lighted by light refracted through the atmosphere on one side of the Earth. (Photograph by the Yerkes Observatory.)

University shows that tidal friction is not limited to the historical period of man but has been in progress since ancient geological periods as far back as the Devonian, nearly 400 million years ago. He finds that certain fossil corals of this period exhibit *annulations*, like tree rings, that measure daily and annual growth rates. Thus the year then contained some 400 days, reducing the day to some 22 hours. C. T. Scrutton of Oxford has since found *monthly* annulations, indicating that the lunar month was then only 21 days in length. We shall return to these extremely important facts in discussing the evolution of the Earth and Moon.

The accurate observations of the Moon, Mercury, Venus, and the Sun during the past century demonstrated strikingly

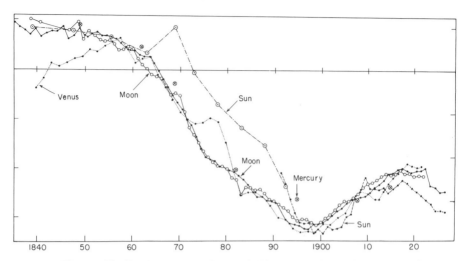

Fig. 83. The Earth is not a perfect clock. The curves above, derived by H. Spencer Jones, represent observed deviations from the calculated motions of the Moon, Sun, Venus, and Mercury. Since these bodies could scarcely deviate by chance in the same fashion, the times of observation must be in error. Therefore the Earth turns irregularly.

irregular variations in the length of the day. In Fig. 83 the deviations in the observed positions of these bodies are reduced to the equivalent time error in the position of the Moon. Since the deviation curves are nearly identical, it must be concluded that the Earth is a bad clock. In the latter part of the nineteenth century, the Earth ran relatively fast by more than a second per year. After 1900 it ran slow by less than a second per year.

A rate of 1 second per year does not particularly exceed the accuracy of the best pendulum clocks and is far below the accuracy of modern atomic clocks. It is a tremendous rate for the Earth considered as a rigid body. On the other hand, if the Earth's radius were to expand or contract uniformly by only a few inches, the observed errors could be explained. There is some evidence that the changes in rate are associated with the occurrence of deep earthquakes. Such an association is not surprising in consideration of the fact that some type of alteration must take place in the Earth to cause its rate of rotation to vary.

In recent years the vibrations of quartz crystals, of atoms in ammonia molecules, of caesium atoms, and of other atoms have been counted and harnessed to operate clocks. The precision attained with such clocks is remarkable and increases continuously with

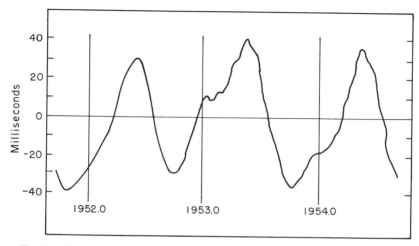

Fig. 84. Seasonal changes in the time of the Earth's daily rotation. (After William Markowitz, U.S. Naval Observatory.)

technological advances. Errors of 0.001 second per year are no longer considered small, while improvements to the extent of another factor of 1000 times are underway. Compared to these clocks the Earth keeps time like a cheap watch. Seasonally on the Earth, the systematic winds, meteorological phenomena, solar tidal changes in the crust, and other minor effects vary the Earth's rate of rotation, producing clock errors like those shown in Fig. 84.

Thus, from the conclusions of this chapter and the preceding, we see that the Earth is a changing dynamic body, affected by forces from within and from without. How different, indeed, is our modern concept of the Earth from that of the ancients!

8

Observing the Moon

As we begin this survey of the Moon's surface, we meet the problem that will confront us repeatedly in planetary studies made
from the Earth—the problem of observing fine detail by means of
a telescope. The telescope is the key of astronomy, an instrument
made with the greatest precision and skill with which astronomers
unlock the closed doors of the universe. But even a perfect telescope
when used on Earth by a most expert observer is limited by an
insuperable handicap, the Earth's atmosphere. This is one reason
that the space program has been so popular with astronomers.

Our knowledge of the surface features of the Moon (Fig. 85) or
of the planets has, until recently, been derived only from a study of
reflected sunlight, which, before orbiting observatories became possible, could reach us only after passing the ocean of atmosphere
above. We have seen that refraction in this atmosphere bends the
light rays through a small angle; unfortunately, no two parts of the
atmosphere refract exactly the same amount. As winds and currents
of warm and cool air circulate overhead, each ray of light is bent
in a slightly different manner. The result is apparent to the naked
eye: the stars twinkle.

Fig. 85. The Moon past first quarter. The south pole is at the top, as it appears in most astronomical telescopes. (Photograph by the Lick Observatory.)

The planets, however, rarely twinkle because they have finite disks; hence they can often be recognized by their steadiness. A telescope magnifies the twinkling so much that often stellar images appear to "boil," as though they were seen across the surface of a hot stove, or across heated desert sands. This boiling turbulence of the atmosphere produces "seeing," which may be relatively good or bad, the quality depending upon the appearance of stellar images in a telescope.

Besides distorting the images of stars, the atmosphere steals away some 30 percent of the incoming light and scatters it in all directions. Above the atmosphere the sky is much blacker at night and equally black in the daytime; thus the stars and planets can be observed by day as well as during the night. On a clear day when Venus is near its brightest it can be seen by the naked eye—if one stands in a shadow and knows exactly where to look. The brightest stars have been reported as visible in the daytime if observed through a tall chimney, a mine shaft, or the like. The reader, however, is advised to use a small telescope or binoculars if he really wants to see stars in the daytime.

Only a few men have had the opportunity of seeing a black sky during a clear day. The first was Major Albert W. Stevens, who from the stratospheric balloon "Explorer II" in 1935 observed the appearance of the sky as seen at an altitude of 13.7 miles above sea level: "The horizon itself was a band of white haze. Above it the sky was light blue, and perhaps 20 or 30 degrees from the horizon it was of the blue color we are accustomed to. But at the highest angle that we could see it, the sky became very dark. I would not say that it was completely black; it was rather a black with the merest suspicion of dark blue . . . To look directly at the sun through one of the portholes was blinding. The sun's rays were unbelievably intense." The astronauts, whose orbits lie completely above the atmosphere, now report similar observations from the vacuum of space.

The brightness of the night sky as seen from the Earth affects visual observations only slightly, because the eye is not sufficiently sensitive to be blinded by such weak light. The photographic plate or image tube, however, can be exposed until it is overwhelmed by the night-sky light. Our atmosphere, therefore, handicaps tremendously the photography of faint nebulous objects, which may be much fainter than the diffuse sky light.

The bad "seeing" sets an insurmountable barrier to observing fine detail, either visually or photographically, on the Moon and planets. Below a certain angular limit neither the eye nor the photographic plate can register any detail. This limit is at best about 0.1 second of arc and corresponds to the theoretical *resolving power* of a telescope with an aperture of 45 inches. The theoretical resolving power varies inversely as the aperture of the telescope and hence becomes 0.5 second of arc for a 9-inch aperture. For bright objects the eye is more effective than the photographic emulsion because it can register details during those rare instants when the "seeing" is nearly perfect. The photographic plate, on the other hand, requires an appreciable exposure time, during which the "seeing" will change. The resolving power of lunar photographs rarely exceeds 1 second of arc, or about a mile. Thus a relatively small telescope, under good "seeing" conditions, can reward the patient observer with an extremely good view of lunar details. Television image tubes, having much greater sensitivity than the photographic emulsion, can sometimes match the eye in detecting fine details.

To minimize the undesirable atmospheric effects, astronomers have searched to "the ends of the Earth," and now *beyond*, to find locations where the "seeing" is exceptionally good. Mountain tops, above the dust and water vapor of lower areas, generally provide very transparent skies, but the "seeing" on a mountain chosen at random may be poorer than at sea level. The Mount Wilson and Lick Observatories, which are responsible for most of the Earth-based photographs of the Moon reproduced in the present chapter, are located on mountain tops in California. The Lowell Observatory was established at an altitude of 7000 feet, on the Flagstaff plateau, Arizona, after Percival Lowell had searched "in Japan; in the Maritime Alps, Algeria, Mexico, California, and Arizona" for the "best procurable air." Similarly, to obtain improved observing conditions the Southern Station of the Harvard Observatory (including telescopes, mountings, and other equipment) was moved from Arequipa, Peru, to Bloemfontein, South Africa. Likewise, the French astronomers developed the 9400-foot Pic du Midi as an improved site for good observations. The best observing sites in the world now appear to be in the Chilean Andes. High-altitude balloons and aircraft are extremely helpful but telescopes in space offer the only ideal solution to the problem of "seeing."

As we have noted, a large telescope will enable an observer to discern finer details than a small telescope—if the "seeing" permits.

When the "seeing" is bad, however, the image of a planet or the Moon in a large telescope may be even poorer than in a small one, because the larger area allows a greater variation of the air conditions. Hence telescopes of moderate aperture (6 to 20 inches) are usually the most effective for direct visual studies. The great reflectors are used almost exclusively with registering devices where their tremendous light-gathering power is of the utmost value. The magnifying power for any telescope is the ratio of the focal length of the objective to that of the eyepiece and may be chosen at will by a change of eyepieces. A high power is used when the "seeing" is good and a lower power when it is poor.

The Moon is a spectacular object as seen through any telescope, whether large or small. Galileo was the first man in history to enjoy this privilege and to record his observations for posterity. Even with his small telescope he could detect the mountains, the craters, and the great dark areas that make up the features of the "man in the Moon." To him the dark areas looked like great seas of water, hence he called them *maria*, the Latin term for seas (singular *mare*, with the accent always on the first syllable).

In Fig. 86, where the Moon is full, the maria can be seen to good advantage. A few of them and other conspicuous lunar features are identified in Fig. 87. The somewhat whimsical Latin names chosen for the maria cannot be explained on any rational basis, although Tranquilitatis, Serenitatis, and Frigoris seem appropriate enough. These maria are, of course, not seas but great plains, fairly flat except for the curvature of the surface, and devoid of both air and moisture.

Mare Imbrium (Sea of Showers) and Mare Serenitatis, in the lower center of Fig. 86 are very large and nearly circular in shape, in so far as they are clearly outlined. The greatest diameter of Imbrium is over 700 miles, and of Serenitatis, 430 miles. A close-up of a part of Mare Imbrium is shown in Fig. 88. The magnificent mountain range outlining the upper left portion of the mare is known as the Apennines. These mountains rise some 18,000 feet above the level of the plain, a height that fully justifies the plagiarism in their name. The perspective given by the shadows in Fig. 88 reveals that the Apennines are great peaks rising sharply from the floor of the mare but sloping away gradually toward the outside. Deep valleys and cuts are numerous. The mountain range presents the appearance of models of terrestrial ranges, where valleys have been worn by the erosion of water—but no water exists on the Moon.

Fig. 86. Rays at full moon. Note the magnificent rays from Tycho (*upper center*) and from Copernicus and Kepler (*middle right*). Note also the many craters that show bright rims. (Photograph by the Lick Observatory.)

Paralleling the inner edge of the Apennines can be seen a long somewhat crooked furrow, or *rill*. Several hundred rills have been found on the Moon. They are ditchlike depressions, hundreds of feet deep and extending for tens of miles along the Moon's surface. Since the rills do not show tributaries, as they should if they

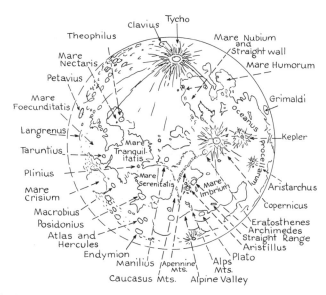

Fig. 87. Map of the Moon, with identifications of certain markings. Compare with the photograph, Fig. 86. (Original drawing by Donald A. MacRae.)

were formed by erosion, and since their walls are not raised above the surrounding "moonscape," they are most rationally explained as cracks. During cooling it is likely that the Moon's surface split open in places. Long or deep cracks may also have been filled in or overflowed by subsurface molten material, and show as a different type of marking. The long low ridge in the upper center of Fig. 88 appears to be an extension of the rill system already noted. A great crack that once opened at the base of the Apennines may now show as a rill along a part of its length and as a ridge in another part, while in between it may have been entirely covered by the flow of molten material.

The more scattered mountains in the lower left of Fig. 88 are the Alps. Their most conspicuous feature is the Alpine Valley, a giant cut through the center of the chain. The Valley is 6 miles wide at its broadest portion and 75 miles long, with a level floor. The magnificent photograph of the Alpine Valley region by the U.S. Lunar Orbiter IV (Fig. 89) shows a central rill in the valley, many other rills, and the irregular nature of the mountain range. There can be little doubt that the rills are the result of a "geological" type of fissuring and partial filling. A very few show branching and may possibly represent some type of flow.

Fig. 88. Mare Imbrium; Moon at third quarter. (Photograph by the Mount Wilson and Palomar Observatories.)

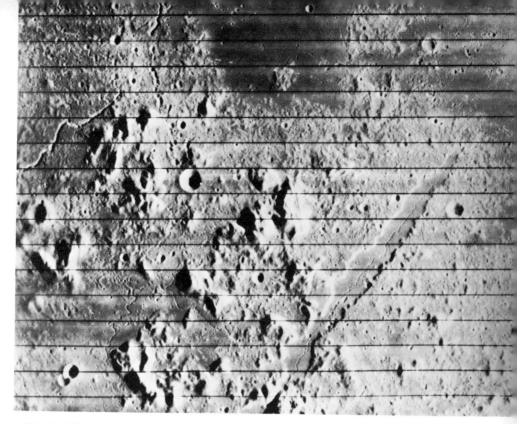

Fig. 89. The region of the Alpine Valley as seen by the U.S. Lunar Orbiter IV. (Courtesy U.S. National Aeronautics and Space Administration.)

The rugged appearance of the Mare Imbrium circumference in Fig. 88, as compared to the general dullness of Fig. 86, is not produced by increased photographic contrast in the first reproduction, nor by an effect of enlargement. In the first view the Moon is full, and the Sun is shining directly down on it; hence the shadows are eliminated. In the second view the Sun is shining from the right, casting long shadows across our line of sight. Because of the curvature of the Moon, the extreme left edge of Fig. 88 is in darkness, except for the high mountain peaks. Along the *terminator,* between darkness and light, the rugged features of the Moon show to the best advantage. The great shadows betray the irregularities that may be invisible when the Sun shines overhead. Because of this effect, the Moon can be best observed when near the first or third quarter. The Sun's rays along the terminator are nearly perpendicular to our line of sight. At full moon, we can distinguish only the light and dark areas; the irregularities are lost.

The shadows serve a very useful purpose, in providing an accurate measure of the heights of the lunar features. In Fig. 90*a* the

Fig. 90a. Piton, an isolated lunar peak in Mare Imbrium; Moon at third quarter. (Section of a photograph by the Lick Observatory.)

isolated mountain peak, Piton, in the lower left area of Mare Imbrium is even further enlarged to accentuate the shadow. The length of such a shadow can be measured, and the angle of the Sun's rays can be calculated from the phase of the Moon and from the known position of the mountain. Fig. 90b illustrates the geometry of solving for the height. The calculations are straightforward, but somewhat involved because of the several angles that must be considered.

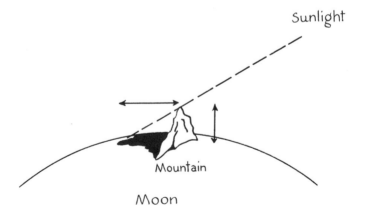

Fig. 90b. A lunar mountain. The lengths of the shadows cast by lunar markings enable astronomers to measure the heights. Compare with Fig. 90a.

Almost everywhere on the Moon there are craters, which become conspicuous when seen near the terminator. The varied character of the craters is apparent in the region of Mare Imbrium, where they stand alone in the open plain. Some seem smooth and flat within, while others show one or more central peaks, often perforated by smaller craters. Crater walls barely rise above the plain here and there, some of them only partially complete. Close examination reveals that the Moon's surface is covered with an almost unlimited number of small craters (craterlets and crater pits), as though the Moon had been peppered with buckshot. The smallest craters identifiable in the Earth-based photographs here are usually a mile or two in diameter.

The craters can be classified into several types according to their forms. Since fine distinctions can be extended interminably and since the definitions are not always concise or uniform, it is perhaps better not to stress the type names. Many craters possess interior plains as flat as the maria and mountain walls that rise abruptly to define the edges. These are known as Bulwark Plains, bulwarked plains, or walled plains. The level of the plain may lie above or below the general level outside. The maria, too, may be depressed or raised with respect to the average surface level. The largest clear-cut crater on our side of the Moon, Clavius, is a bulwarked plain, with a maximum diameter of 146 miles from opposing mountain summits. Clavius can be seen near the top center of Fig. 91. The curvature of the Moon's surface is sufficient to hide the 20,000-foot mountain walls from an observer standing in the center of the plain.

The crater Tycho, seen slightly above the center of Fig. 91, represents a somewhat different type of crater formation, often called ring mountains. Only a small fraction of the basin is flat, the crater being more nearly saucer-shaped. The inner slope of the mountain rim is itself ringed, somewhat as though laminated or terraced. These ring-mountain craters are also fairly perfect in form, almost circular, and are rarely encroached upon by lesser craters or other deformations. The ring mountains thus show evidence of having been formed later in the Moon's history than the bulwarked plains, which bear the scars of subsequent tribulations. Other fine examples of the ring mountains are Eratosthenes and Copernicus (Fig. 88, upper right, and Figs. 92 and 93).

The rougher areas of the Moon (Fig. 91) are completely covered with a wild hodgepodge of craters within craters and craters upon craters. They all appear to have been formed in an entirely hit-or-

Fig. 91. Lunar bad lands. The southern area of the Moon at third quarter. (Photograph by the Mount Wilson and Palomar Observatories.)

Fig. 92. Copernicus. Moon past third quarter. Compare with Figs. 86, 93, and 94. (Photograph by the 100-inch reflector of the Mount Wilson and Palomar Observatories.)

miss fashion, the newer ones evolving with a complete disregard for all that were there before. Sections of the wall may stand after an old crater has been partly demolished by a new one, and this, in turn, may be pock-marked by smaller craters still more recent.

Across these rugged areas of the Moon and across the extensive

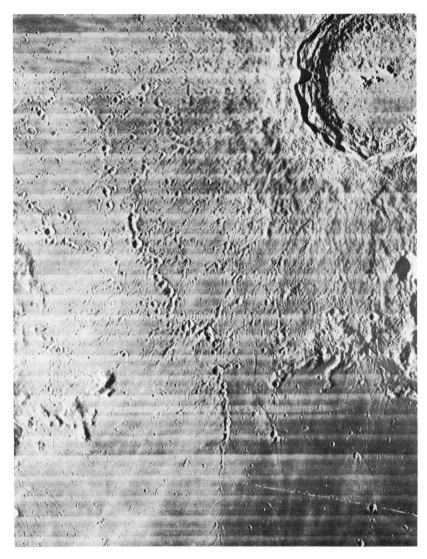

Fig. 93. Part of the Copernicus area, photographed by U.S. Lunar Orbiter IV. (Courtesy U.S. National Aeronautics and Space Administration.)

plains run great systems of *rays*, the light-colored streaks that show so conspicuously when the Moon is full (Fig. 86) but almost disappear at the partial phases (Fig. 85). A most notable system centers on the crater Tycho (diameter 54 miles), from which the rays can be traced almost around the Moon. In Fig. 91 the rays are barely discernible, and Tycho has become just one of many craters, not the

Fig. 94. The region east of Copernicus. (Courtesy of G. P. Kuiper.)

most conspicuous of all. The rays cast no shadows and can be detected only by their lighter coloring. They are broken neither by mountains nor by any other features of the lunar topography. They are certainly splash marks from large, relatively new craters.

Note the complex structures about the craters Copernicus and Kepler in Fig. 86. The light color of the rays is shared by the craters with which they are associated, and also distinguishes the rims of a large number of craters. At first glance the region about

Fig. 95. The Straight Wall throws a dark shadow when illuminated from the west (*left*). Compare Fig. 91. (Section of a photograph by the Yerkes Observatory.)

Copernicus in Fig. 86, when the Moon is full, can hardly be identified with the area about Copernicus in Fig. 92, at a later phase. The bright crater rims in the latter figure, however, can soon be detected in the former, while some of the larger craters with dull edges have faded to invisibility. A careful comparison of the two photographs is most instructive. The second largest crater in Fig. 92, Eratosthenes, can hardly be found in Fig. 86, although it is very

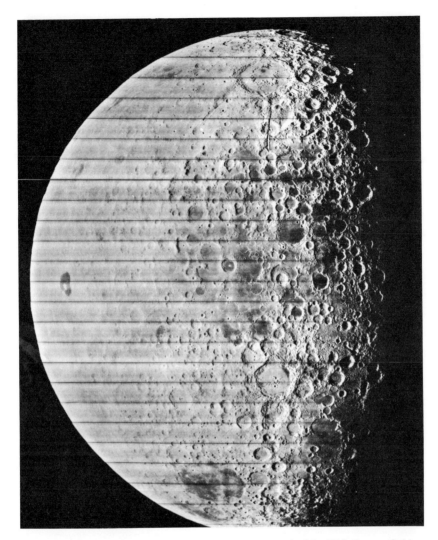

Fig. 96. Part of the far side of the Moon, photographed by U.S. Lunar Orbiter IV, centered near the eastern limb as seen from the Earth and showing the South Pole at the top. (Courtesy U.S. National Aeronautics and Space Administration.)

similar to Copernicus in type. Copernicus, Eratosthenes, and Kepler are 57, 36, and 20 miles in diameter respectively.

To the left of Copernicus in Figs. 92 and 93 are several crater chains, forming a pattern. Such crater chains are not uncommon on the Moon's surface. They are especially noticeable when viewed with a high magnification under good seeing conditions. The highly

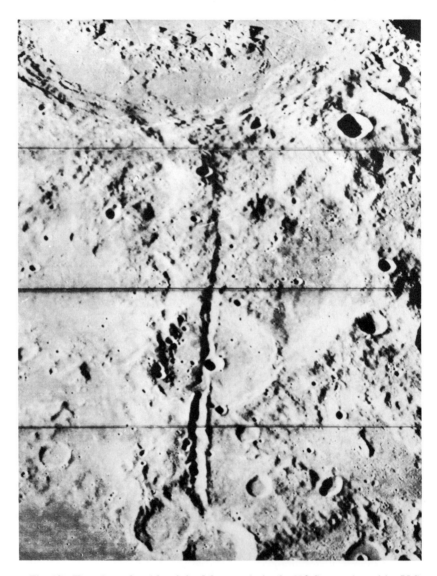

Fig. 97. Trough on far side of the Moon at latitude 65° So. as viewed by U.S. Lunar Orbiter IV. Somewhat like the Alpine Valley, this feature is 150 miles long. Compare Figs. 88 and 89. (Courtesy U.S. National Aeronautics and Space Administration.)

magnified section of regions near Copernicus in Figs. 93 and 94 show formations that appear to be *volcanic domes* and *volcanic sinks*. Some of the domes have central craters and resemble terrestrial volcanoes of the Vesuvius type.

Long serpentine ridges can be seen in the flat plains of Fig. 95 and also in Mare Imbrium (Fig. 88). These low ridges are obviously flow markings, undoubtedly formed by distortions of the area, possibly when the floors of the maria were in a viscous condition. On a flat surface of tar, a horizontal distortion will produce an identical type of marking, especially if the tar is slightly warmed to flatten out the rough ridges.

Another interesting lunar formation is shown in Fig. 95, the Straight Wall (or the Railway) some 70 miles in length, which appears also in Fig. 91. In this picture, the sun is shining from the right and the Straight Wall shows as a *white* line (to its right, nearly parallel to it is a shorter curved rill). The photograph of Fig. 95 was made just after first quarter so that the Sun shines from the left. The Straight Wall now shows as a *dark* line, which proves that the marking is a long straight cliff or wall facing toward the right. Measurement shows that much of it is elevated some 2000 feet above the plain. It is not as steep as we might suspect, however. J. Ashbrook finds that its slope is not more than 40°; it is clearly a rock *fault,* where one edge has risen above the other. Violent "moonquakes," in the dim and distant past, were probably associated with this and other walls visible on the Moon. Similar fault markings, frequent but smaller on the Earth, are the foci of earthquakes. On the Moon, large systems of parallel cracks and faults can be traced, indicating that the surface has been strained and distorted by internal forces.

The other side of the Moon no longer remains a mystery. On October 4, 1959, the U.S.S.R. first achieved the remarkable feat of obtaining images of the Moon from a "cosmic rocket" passing beyond the Moon. Since then the U.S. Orbiter spacecraft sent around the Moon by the Langley Research Center of the National Aeronautics and Space Administration have surveyed almost the entire lunar surface with resulting photographs of extraordinary quality and detail. Figures 96 and 97 show the far side of the Moon and a great gash on it 150 miles long

The major scientific surprise given us by the far side of the Moon is the lack of maria. Whereas the near side is about half covered with maria, the far side shows only one relatively small one! The speculations aroused by this observation are still not satisfied by any complete and acceptable scientific explanation.

There are so many interesting formations on the Moon's surface, individual craters with unusual structures, peculiar rills, rays, maria, mountains, and cliffs, that descriptions could be continued indefinitely. The reader, however, may wish to do some exploring himself, by means of these photographs or with almost any telescope that can be firmly supported. The huge collections of photographs obtained by the National Aeronautics and Space Administration's Ranger, Surveyor, and Orbiter spacecraft offer a lifetime of detailed study of the Moon. Larger-scale charts can be obtained easily by anyone who wishes to learn the proper names of individual lunar formations. Some 5000 markings are officially designated by the International Astronomical Union, and many thousands more have been plotted by assiduous selenographers. The close-up spacecraft photographs are now taxing the official map makers with a job comparable to that of mapping the Earth itself.

The more serious observer can join the ranks of those dedicated amateurs who have contributed so much to solar-system astronomy. In most countries he can find groups who are engaged in patrolling the lunar and planetary surfaces for changes that the professional astronomer will probably miss. Even with the modern acceleration of observational astronomy, there are far too few professionals for the task at the telescope. Any reader who does not know where to turn for local advice is encouraged to write the author for information about observing needs, the groups nearest to him who are engaged in observing artificial satellites, lunar phenomena, occulations of stars by the Moon, planetary surfaces, variable stars, meteors, comets, or other sky activities.

9

The Nature of the Moon

The Moon's surface, as we have viewed it in the previous chapter, is a sublime desolation. The lunar plains are more barren than rocky deserts. The lunar mountains are more austere than terrestrial peaks above the timber line. Lava beds of extinct volcanoes are more inviting than the lunar craters. There is no weather on the Moon. Where there is no air there can be no clouds, no rain, no sound. Within a dark lunar cave there would be eternal silence and inaction excepting possible moonquakes. A spider web across a dim recess in such a cave would remain perfect and unchanged for a million years.

There are no colors in the Moon's sky, only blackness and stars during the bitter night, 2 weeks in length, and then the glaring Sun during the equally long day. But there is danger to man on the Moon's surface. Meteoritic dust may puncture his space suit or his pressurized dwelling and may splash lunar material at rifle speeds to accomplish the same end. Cosmic rays and high-energy particles from the Sun strike him directly because there is no atmosphere to stop them.

If any permanent changes have occurred in the lunar landscape during the centuries of intensive telescopic observation, the changes are too small or uncertain for the observers to be able to agree upon their reality. However, a number of observers have noted haziness, brightening, or reddish coloration in certain crater floors and around the edges of certain craters and maria. Such evidence, largely unsupported by simultaneous observers or photographs, had received little scientific credence until the observations made by N. A. Kozyref at the Kharkov Observatory in the U.S.S.R. On November 3, 1958, he was guiding the slit of his spectrograph (*see* p. 163 for description) on the central peak of Crater Alphonsus (diameter 73 miles; Fig. 98). He first observed that the peak "became strongly washed out and of an unusual reddish hue." Two hours later he "was struck by its unusual brightness and whiteness at the time." His spectrograms confirmed his visual observations, showing in the first case that the light of the central peak was considerably weakened in the violet, compared with the neighboring details of the crater, and in the second that bright bands showed in the light from the central peak. It appears that gas effusion from the crater lasted for not less than half an hour and not more than 2½ hours that night. These and other observations indicate strongly that on rare occasions appreciable quantities of gases are emitted from cracks or craters in the lunar surface. Possibly there are occasional morning "fog" effects in certain of the craters.

Since Kozyref's observations, various systematic visual and some photographic programs have been initiated to look for these changes. B. M. Middlehurst and P. A. Moore have catalogued some 400 similar reports, including one nearly two centuries old by William Herschel; many of these are uncertain or easily misinterpreted. Middlehurst and Moore conclude that these transient phenomena usually last for a few minutes and appear around the peripheries of maria, ring plains with dark floors, and about ray craters. None appear on the lunar highlands. Furthermore, the reported brightenings do not correlate with solar activity, but are partially related to lunar perigee and apogee. Can there be gas emanations stirred up by tidal strains in the Moon's surface layers, maximum when the lunar tidal deviations are greatest?

There is evidence for florescent light on the lunar surface arising possibly from solar ultraviolet light and high-energy particles in space. It shows best on sunrise areas of the Moon where the solar light probably causes the surface layer to radiate the energy stored up by impacts with ions of the solar wind.

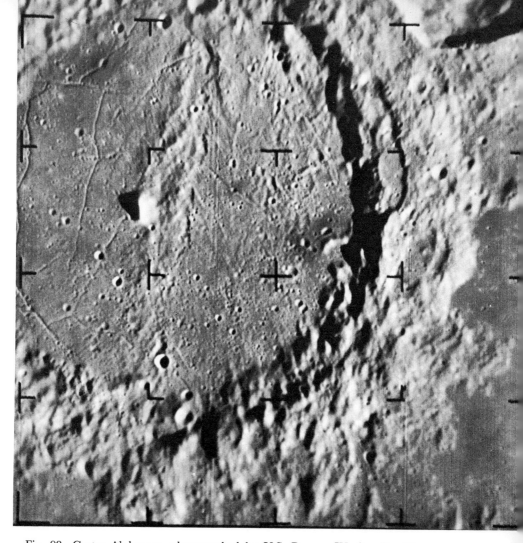

Fig. 98. Crater Alphonsus, photographed by U.S. Ranger IX. See Fig. 91, second crater from bottom, left center. (Courtesy U.S. National Aeronautics and Space Administration.)

Beyond these observations there is no evidence to show that the Moon has any atmosphere. Also there is no evidence of erosion by either wind or rain. Over the years more subtle observations of the Moon have continuously pushed down the upper limits to the possible atmosphere on the Moon. Lack of refraction of stars near the Moon's surface and lack of a visible aurora to be expected in a residual atmosphere both show that the Moon's atmosphere must be extremely rare. The most sensitive measurement has concerned the absorption of radio noise from the Crab Nebula as this "radio star" was occulted by the Moon. B. Elsmore has placed the limit of the lunar atmosphere as less than 2×10^{-13} (that is, less than a million-millionth) of the Earth's atmospheric density near sea level.

We should suppose the Moon to be devoid of an appreciable atmosphere because of the smallness of its mass. Its surface gravity is insufficient to prevent the molecules of an atmosphere from being hurled into empty space. Any body, large or small, moving away from the Moon's surface with a speed in excess of 1.5 miles per second would continue to recede indefinitely, completely out of the gravitational control of the Moon. This critical velocity of escape is only slightly greater than the *average* speed of a hydrogen molecule in a gas at ordinary temperatures. Since some of the molecules must always move faster than the average, a hydrogen atmosphere would dissipate from the Moon almost instantly. The dissipation of oxygen or nitrogen would be very much slower because the molecules are heavier than those of hydrogen. In a short time astronomically, however, the Moon would lose any atmosphere it might once acquire. Quite probably the Moon now gains some atmosphere from interplanetary space, mostly from the gas clouds shot out by the Sun and the gas lost from the solar corona. This extraordinarily tenuous atmosphere is continuously lost, while the high energy of the incoming atoms probably knocks away the heavier atoms that may be oozing out of the Moon's interior.

Among the active forces that are probably most important today in altering the Moon's surface are meteoritic impacts and the high-energy particles, particularly the x-rays and the cosmic rays, that penetrate the top surface layers. None of these forces may produce much change during a human lifetime, but over hundreds of millions of years they have appreciably altered the Moon's surface. Man himself will probably make striking alterations in the Moon's appearance, but they are yet hard to find. In February 1967 the U.S. Lunar Orbiter III photographed the Surveyor I spacecraft that soft-landed, June 2, 1966, on a typical "smooth" mare area in Oceanus Procellarum. The picture shows the white spot of the spacecraft and its shadow some 30 feet in length. The U.S. Ranger VIII spacecraft crash landing produced a crater some 40 feet in diameter, as identified in an Orbiter photograph. Several other crash landings on the Moon undoubtedly produced similar craters.

The rate of meteoritic influx on the Moon is estimated at perhaps a foot per billion years. But the Moon probably loses more mass than it gains by this accretion of material from space. Each meteorite striking the Moon produces a miniature explosion, throwing rocky and meteoritic materials out in all directions. At a speed of 20 miles per second the particle can impart a velocity greater than

1½ miles per second, the velocity of escape from the Moon, to more than its own mass of material. Since there is no atmosphere to resist the loss, such material leaves the Moon's gravitational field although much may be recaptured from Earth orbits.

The alternations in temperature from the lunar noon to midnight are extreme, more than 400 Fahrenheit degrees (Fig. 99), but they take place slowly. The *exfoliation* or flaking away of the surfaces of terrestrial rocks is due chiefly to the expansion of absorbed moisture on freezing. Because of the absence of water on the Moon, exfoliation could result only from thermal expansion and contraction. However, the very gradual changes in temperature on the Moon allow the rocks sufficient time to adjust their internal temperatures that exfoliation by pure expansion and contraction is very slow.

Although the enormous temperature changes from day to night occur very slowly, large changes are observed during an eclipse of the Moon. E. Pettit and S. B. Nicholson of the Mount Wilson Observatory first measured the Moon's temperature by its infrared (or heat) radiation through the course of a lunar eclipse. The temperature fell from $+160°F$ to $-110°F$ in about an hour. Such a quick alternation in temperature may be more active in exfoliating newly exposed lunar rocks, but only on the near side of the Moon. The high-energy particles of the solar wind etch the lunar surface very slowly and possibly darken it. They and cosmic

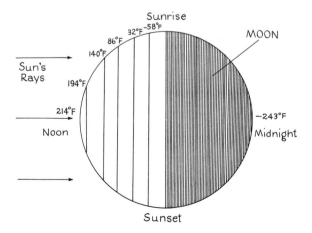

Fig. 99. Temperatures on the Moon. (From measures by E. Pettit and S. B. Nicholson.)

rays deteriorate slightly any crystalline structure at the very surface of the Moon.

Had the Earth been unable to retain an atmosphere, its surface would probably be similar to that of the Moon, barren and probably rougher. Air and water, however, have made possible the surface that we know and have led to the evolution of living forms. No life can exist on the surface of the Moon, although conceivably there may have been a time during its formation when there was appreciable atmosphere. Primitive microscopic life forms possibly may then have developed, but this possibility is still speculative.

Radio telescopes have added enormously to our knowledge of the Moon's distance (Figs. 100 and 101) and temperature. Active radar pulses can be timed in transit to give the distance and also the roughness of the lunar surface. Passive radio receivers, on the other hand, can measure the radiation emitted by a surface and thus determine its temperature. All ordinary materials radiate more energy at all frequencies when they are heated. From earthy materials some distance below the surface radio waves may pass through the porous upper layers to give an indication of the temperature beneath. At wavelengths of 30 centimeters or longer there is practically no temperature change observed during the month or even during lunar eclipses, while at successively shorter wavelengths, nearer to the infrared, temperature changes are increasingly greater. A few feet below the Moon's surface on the equator, the temperature appears to remain constant at $-60°$ F or perhaps somewhat colder.

These combined results show that the Moon's top surface layers are made of very porous material, an excellent insulator and probably quite thin. This insulating layer holds little heat and thus can respond quickly to changes in solar radiation, at sunset, sunrise, or eclipses of the Moon. Infrared measures over the near surface of the Moon show that many areas remain warmer during eclipses and in the lunar "evening" than does the average lunar surface. Figure 102 shows a map of these regions, which comprise almost all the new (white ray) craters such as Tycho and some of the maria. The lunar highlands are relatively free of such "hot" spots except for new craters.

We can only conclude that more recently disturbed areas of the Moon have not had time to develop so thick an insulating surface as the older areas. Thus the heat accumulates during the lunar day

Fig. 100. First radar contact with the Moon. At a wavelength of about 1.5 meters in January 1946 this antenna at the U.S. Signal Corps Engineering Laboratory, Fort Monmouth, N.J., first bounced radar pulses (Fig. 101) off the Moon. (U.S. Army photograph.)

in rocks or good conducting materials, to be radiated away during the night. Recall the warmth from a wall in the early evening after it has been exposed to the Sun during the afternoon. The observations do not yet clearly indicate whether any areas on the Moon actually radiate heat from an internal source, say of a vol-

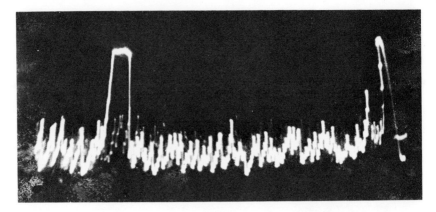

Fig. 101. First radar pulse from the Moon. The radar antenna is shown in Fig. 100. (U.S. Army photograph.)

canic nature. Very probably some heat sources will be found, however, as the longer wavelength radio measures suggest a slight excess of heat from the interior.

Since dust in a vacuum is an excellent insulator to heat, the rapid temperature changes during eclipses have long suggested that the upper layer of the Moon's surface might be dust. But one should not expect *loose* dust, as has been suggested by one astronomer, because dust particles tend to stick together in a vacuum and because there is a little "cement" provided by gas generated from high-velocity atoms and meteoric encounters with the Moon.

The great increase of the Moon's brightness near the full phase indicates that the surface is extraordinarily rough at dimensions of less than a millimeter. The total light of the Moon doubles in the 2 days before full Moon and the surface appears nearly uniformly bright to the edge at full. But long-wavelength radar observations of the Moon show that its surface is fairly smooth, with about half of its area tilted at more than 8° from the spherical. From limited data the maria appear smoother than the craters and mountains. At a wavelength of 8 millimeters, radar measurements made at the Lincoln Laboratory of the Massachusetts Institute of Technology show that the Moon's surface is much rougher than at longer wavelengths. The central area of reflection is much larger. In visual and infrared light, however, the full Moon appears almost uniformly illuminated.

The great triumph of spacecraft actually soft-landing on the Moon now gives us as clear a picture of the lunar surface as an

Fig. 102. Infrared picture of Moon during eclipse shows white spots that cool more slowly than remainder of surface. Compare Figs. 86 and 87 to identify major ray craters and certain maria. (Courtesy R. W. Shorthill and J. M. Saari, Boeing Aircraft Co.)

actual close-up view by the eye. The U.S.S.R. Luna 9 landed on January 31, 1966 (Fig. 103), the U.S. Surveyors I (Fig. 104) and II on May 30 and September 20, 1966, the Luna 13 on December 21, 1966, and the Surveyor III, with its extensible shovel, on April 20, 1967. All, particularly those of the shovel and the imprints of the Surveyor pads (Figs. 105 and 106), give invaluable information about the lunar surface. The top material is mostly very fine grained, the grains being on the order of a few thousandths of an inch in dimension, interspersed with occasional rocks from pebble-size upwards. The bearing strength is a few pounds per square inch, so that a man walking on the surface would leave clear-cut

Fig. 103. The first close-up of the Moon. From the U.S.S.R. Luna 9. The largest nearby rock is about a foot in size.

footprints a fraction of an inch deep. They would remain visible perhaps a million years.

The shovel and pad leave rather vertical walls in their cuts through the surface, indicating that the material has internal strength, but is weak, perhaps a little stronger than freshly plowed earth. The surface when gouged tends to crack slowly in small blocks, confirming that it is a somewhat "crunchy" material, as the author had predicted. In other words it is not very compressible under moderate pressures, suggesting an internal structure of some coherence. The surface density, best determined by Luna 13, is near the density of water. It must be rather porous and becomes denser and stronger at depths over the few inches so far studied.

Thus the optical, radar, and close-up data show that the Moon is extremely rough over very small distances and becomes relatively smooth over distances of a few inches, except where occasional rocks protrude or lie on the surface. These appear to have been thrown from crater explosions. Large-scale steep slopes on the Moon are rare except in craters. Even the Straight Wall has a typical slope

Fig. 104. Mosaic view of lunar surface near U.S. Surveyor I. Note footpad (diameter 12 inches), white rocks, and craters. (Courtesy U.S. National Aeronautics and Space Administration.)

of only 40°; the average slopes of the inner walls of the crater Tycho are only about 17°. The U.S. Ranger spacecraft pictures of the Moon show that only one percent of the surface is inclined more than 13°. The great shadows seen near the terminator deceive us as to the ruggedness of lunar features. Mount Piton, for example (Fig. 90a), rising some 7000 feet above the mare floor, stretches out more than 70,000 feet at its base and has a level top.

But one major puzzle of the lunar surface remains, even after the close-up views from the soft-landings. Why is the surface so very dark? Fine dust from broken rocks is a much better light reflector than the Moon's surface. The trench made by the shovel of Surveyor III shows that the very top layer (less than a millimeter thick?) is slightly lighter in color than the deeper material. But why is the deeper material so very dark and why are some of the rocks so much whiter? For a while it was thought, from laboratory experiments, that the high-energy ions of the solar wind would slowly darken the surface. But later evidence shows that this effect may have arisen by contamination, perhaps by carbon atoms.

The occurrence of occasional whiter rocks (Albedo ~0.2) in the Surveyor pictures suggests that the material deep in the Moon is whiter, the rocks having been excavated and thrown great distances by large meteoritic impacts. Has the surface layer been darkened chemically in lava or lava dust flows? Is it made of organic matter left over from the Moon's early days when it may have had an atmosphere and even water? Or is it just very dark lava? We still have no knowledge of the sub-surface layers on the highlands, as all

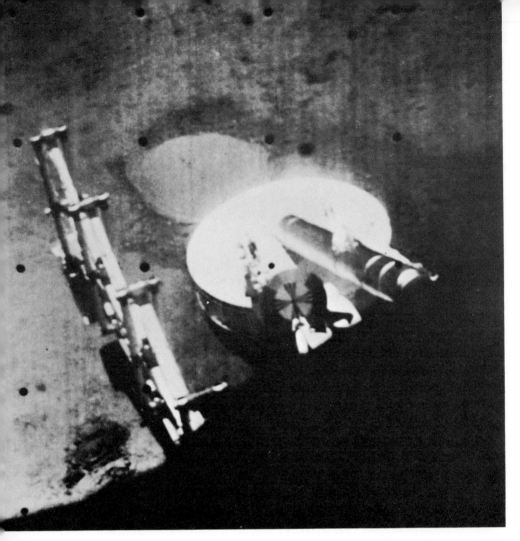

Fig. 105. Imprint of U.S. Surveyor III footpad on surface of the Moon. Black dots are fiduciary marks. (Courtesy U.S. National Aeronautics and Space Administration.)

of the soft-landings made thus far have been planned for the maria. The highlands are brighter than the maria so perhaps their underlying material is more like that of the white rocks, but their surface has been darkened somewhat.

Some small craters show an inner terrace or even two, suggesting that a harder stratum or strata lie a few meters to tens of meters below the surface. Such layering of the lava flows is evident in several of the maria regions that are well studied. In addition, in these regions of the maria the craters become *saturated* below diameters of perhaps fifty to a few hundred feet. That is, new

Fig. 106. U.S. Surveyor III shovel and trench on Moon. The width of the claw is two inches. (Courtesy U.S. National Aeronautics and Space Administration.)

craters have overlapped and covered up old craters to produce a statistically stable distribution of craters. The addition of new craters would not change the general appearance of the surface

The Ranger and the Orbiter pictures all show that some of the sharply defined "new" craters have white rims and some have rims of the same shade as their surrounding. Hyperballistic experts suspect that this difference arises from the velocity of meteoritic impact. At low velocities the meteorite causes a subsonic explosion and the broken material is merely pushed out. But at velocities of 10 miles per second or higher, typical of meteorites from space, much of the

Fig. 107. A 500-foot diameter crater strewn with 3-foot boulders as viewed by U.S. Lunar Orbiter III in Oceanus Procellarum. (Courtesy U.S. National Aeronautics and Space Administration.)

material would be pulverized into very fine dust and thus give a white rim. Figure 107 shows a very rocky crater wall and surroundings, probably caused by a low-velocity impact.

The huge rays from the great new craters such as Tycho cannot, however, be explained by white dust alone. The U.S. Ranger VII pictures have confirmed Kuiper's telescopic observation that the rays are rough and rocky. White rocks, such as appear in the Surveyor pictures (Fig. 104), could cover the surface of the rays sufficiently to keep them relatively white for long periods of time, until they were slowly covered by debris thrown from more distant parts of the Moon. Their increase in relative brightness at full Moon, however, requires further explanation. Possibly their very roughness produces the effect. The "shadow" of an aircraft at great altitudes becomes a bright spot on ordinary ground surfaces although it is a dark shadow on water (see Fig. 134). On foliage, for example, light rays opposite the Sun can retrace their original path to the eye. At other angles the light is dimmed in finding a new path. Thus if the lunar bright rays are rougher on a large scale than the surface of the Moon on a microscopic scale, they then show their true higher albedo near the full phase.

For many years before the space age there was great speculation about the origin of the Moon's surface features. Has the Moon duplicated the volcanic or plutonic activity of the Earth with great volcanic craters and lava flows? Or was it always rather inactive with meteoritic impacts being the great surface molding agent? Was the Moon ever molten or partially molten?

We look to the Moon's surface itself for visual evidence concerning melting and volcanic action. We note in many areas of the Moon, particularly on the far side, a chaotic maze of crater upon crater, constituting the rough "highlands" area of the Moon. In contrast the surfaces of the maria appear relatively smooth and contain many fewer intermediate-sized craters per unit area. There, however, we find abundant evidence for volcanic or plutonic activity: the crater chains near the great crater Copernicus, mentioned earlier (Fig. 92), and in a region east of Copernicus where volcanic sinks and domes also appear (Fig. 93 and 94). The remarkable crater Wargentin (Fig. 108) shows clearly that lava was actually forced to the top of the crater walls, leaving it essentially full, or that regions nearby later subsided from a common level. Volcanoes generally cause a mountain of material to rise about a central vent, as in Vesuvius, the volcanoes of the Hawaiian Islands, and many others (Fig. 109). Lunar evidence for volcanoes of these dimensions can be seen in Figs. 93 and 94, but all of these clear-cut cases of volcanoes and volcanic domes are relatively small in size.

Cracking, melting, slumping, filling, ridging, and plutonic activity of many kinds are apparent almost universally not only on the maria, but within the filled craters and in flat regions between

Fig. 108. The crater Wargentin. The two largest craters are Schickard (*lower*) and Phocylides (*upper*), with Wargentin, the filled crater, between them to the right. (Photograph by the 120-inch reflector of the Lick Observatory.)

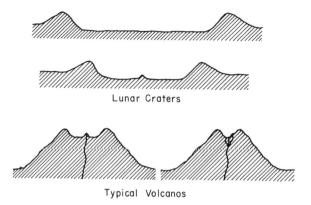

Fig 109. Craters and volcanoes. These schematic cross sections indicate the basic differences between lunar craters and volcanoes of the Vesuvius type.

lunar mountains. For example, the U.S. Ranger IX views within the crater Alphonsus (Fig. 110) show that the rill system and most of the small craters must have arisen from a geological type of activity. Dark halo and slump craters are abundant. The U.S. Orbiter II view of Oceanus Procellarum in the neighborhood of Crater Marius (Fig. 111) shows not only serpentine ridges that certainly were formed by compression lava or magma flows, but also volcanic domes of diameters 2-10 miles and heights 1000 to 1500 feet, some with visible craters. The remarkable complexity of the maria surfaces seen close up, as compared to their drabness in Earth-based photographs, attests to the plutonic activity of the Moon at some yet-unknown period in the past.

On the other hand, the larger craters on the Moon are all or mostly meteoritic. Ralph B. Baldwin in 1949 amassed such an array of evidence for the meteoritic theory that very few proponents of the volcanic theory now remain. The number of new craters on the maria fits well with the expected influx of large meteorites over 2 to 4 × 10⁹ years. There was the old counter argument that, if such large meteoritic bodies made the great craters on the Moon, the Earth should carry even greater scars because of its increased gravity and the consequent greater velocity of fall of similar bodies to the Earth. In the last few decades increasing evidence firmly proves that the Earth *has* many scars of this type, in the form of crypto-volcanoes, not to mention the recognizable recent meteor craters such as the Barringer Crater in Arizona and larger ones well demonstrated by C. S. Beals in Canada (Fig. 53). Geological forces

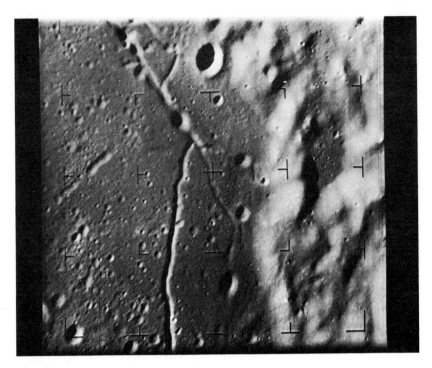

Fig. 110. Rill system inside the crater Alphonsus. Note the dark halo crater, top center, which may have been darkened by gas ejection. (Courtesy U.S. National Aeronautics and Space Administration.)

have, of course, filled large craters of the past, tilted them, eroded them away, and in large measure destroyed their record. A striking example is the Vredefort Dom in South Africa, which was originally more than 30 miles in diameter. A discussion of these fossil meteoritic craters on the Earth is found in *Between the Planets* by F. G. Watson.

We know from our experience with nuclear weapons that large craters can be formed by explosions. High-velocity meteorites can produce similar effects on *any* scale. E. M. Shoemaker's diagram in Fig. 112 shows the impact craters in the area around Copernicus (compare with Figs. 92 and 93). These tiny craters, still of the order of a mile in diameter, have been produced by the material thrown from this explosion. No volcanic activity that we know of could produce such large explosive action as to throw out masses of millions of tons of material to distances of 10 miles and much farther. On the other hand, craters of volcanic type do appear in parts of this same area (Figs. 93 and 94).

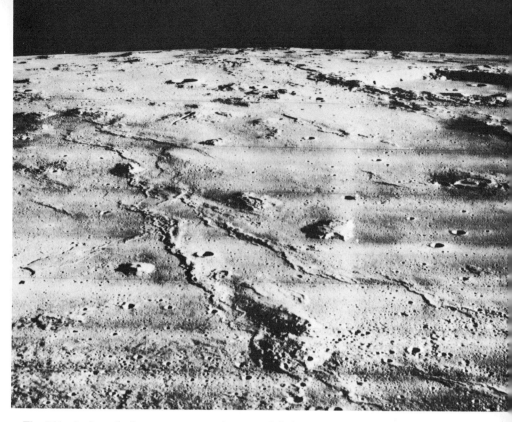

Fig. 111. A view of a lunar mare near the crater Marius, upper right, which is 25 miles in diameter and a mile deep. (Courtesy U.S. National Aeronautics and Space Administration.)

Other evidence that the more perfectly formed and presumably newer lunar craters are of meteoric origin comes from *Schröter's rule,* discovered by the selenographer J. H. Schröter (1745–1816). He found that some craters have piled up around them just sufficient material to fill them. Thus these craters were formed without appreciable addition or loss of material from the interior of the Moon. In Fig. 109 this characteristic is indicated schematically.

Many lunar craters show a central mountain, much lower than the rim of the crater, oftentimes with a centrally located crater. It now seems possible, as G. P. Kuiper suggests, that these central mountains occur only in craters formed near the critical time when the Moon was partially melted, so that the center of the meteoritic impact could tap fluids to make a central mountain that was essentially volcanic in origin. Craters formed before and after this critical event should not, then, show central mountains. Probably many of the larger craters were finally filled slowly, by isostatic adjustment. During this process internal pressures deep in the crater could have forced up any magma present to form the central peak or moun-

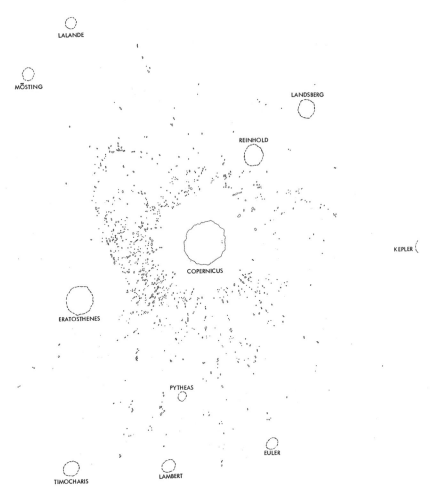

Fig. 112. Secondary impact craters in the area of Copernicus; compare with Fig. 94. (After E. M. Shoemaker.)

tains. The remarkable U.S. Orbiter III horizontal view of the Copernicus crater basin (Fig. 113) and north walls shows again the complexity of plutonic processes after the meteoritic explosion.

Let us now face the question whether the maria were formed primarily by volcanic action or were first triggered by the impacts of extraordinarily large meteorites. The latter solution was first suggested and defended late in the last century by the great geologist G. K. Gilbert (1843–1918), supported by Baldwin in 1949, and more recently by Urey and Kuiper.

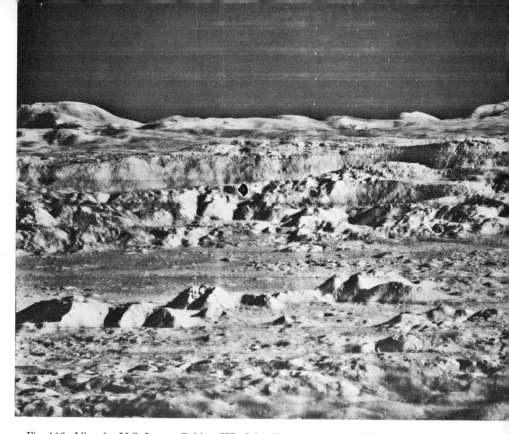

Fig. 113. View by U.S. Lunar Orbiter III of the Copernicus crater. The mountains just beyond the foreground can be seen as the central peaks in Figs. 92 and 93, looking from the south, i.e., from the top of the figures. (Courtesy U.S. National Aeronautics and Space Administration.)

An intense study of the lunar surface shows that at least one-half of the visible surface of the Moon exhibits markings that are linked to Mare Imbrium. A large number of cracks and filled valleys point radially to a point near the center of Mare Imbrium. There are also systems of cracks, mountain chains, and other formations almost circularly symmetric about this point. Figure 114, due to Kuiper, shows this system on a projection normal to the Moon's surface so that the foreshortening of Mare Imbrium, as seen from the Earth, has been eliminated. He identifies the remarkable inner basin surrounded by serpentine ridges marked in dark. This is a nearly square area symmetric about the presumed impact point. The Alpine Valley points directly toward this central point and the great mountain ranges around are quite symmetric with respect to it. Within these mountain ranges is the outer basin in which there appear to be great blocks that represent either original lunar material not destroyed by the formation of the mare or material forced out into that region in the formation of the mare.

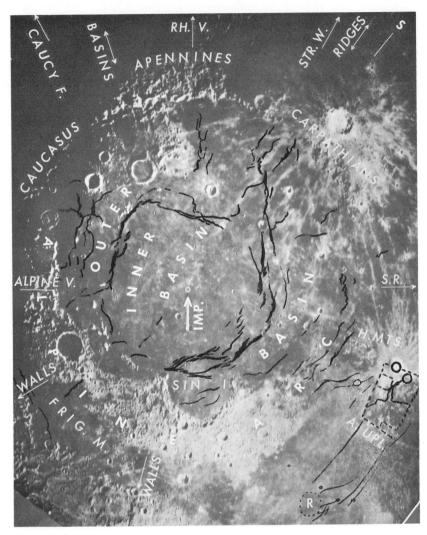

Fig. 114. Structures in Mare Imbrium as delineated by G. P. Kuiper.

Kuiper argues, I think correctly, that an enormous meteoritic body landed near this impact point with a large component of its velocity toward the south (up in the photograph). Within the inner basin a huge crater was formed and materials were thrown out to extreme distances, far beyond the apparent center of the Moon as seen from the Earth. The great impact occurred during the interval when the Moon's outer layers were partially molten, so that the huge mass of material thrown out from the central crater presented a tremendous load for the lunar crustal rocks to support. The present region between the outer basin and the inner edges of the mountain ranges failed to support the load and subsided. The inner edges of the Alpine Mountains, the Caucasus Mountains, the Apennines, and the Carpathians are thus slip faults formed as the inner and initially higher regions sank. Even though some of these mountains are more than 5 miles high, we note that lunar gravity is only one-sixth that of the Earth. Thus the highest mountains on the Moon require a supporting force equal only to mountains less than 1 mile high on the Earth. Hence, if only a small fraction of the upper layers of the Moon were molten, it seems reasonable that enough support could be found from underlying and undisturbed lighter rocks for such a broken and incomplete ring of mountains as now remains. Perhaps too, the ejected material was less dense than the deep underlying material so that isostasy is now maintained.

The region on the Moon near its apparent center, extending to the Apennine Mountains and westward toward the Haemus Mountains near Mare Tranquilitatis, appears muddy and poorly resolved even in the best observing conditions. It is difficult to explain this unusual appearance except by assuming that the area had been covered to a considerable depth by partially melted material thrown radially from Mare Imbrium. This is the basis of Kuiper's argument that the incoming meteoritic body was moving in a southerly direction.

The space studies of the Moon reveal an even more spectacular basin than Mare Imbrium. This feature, Mare Orientale (Fig. 115), has long been partially visible at the Moon's limb, but its true appearance could only be perceived from space. Its outer scarp, the Cordillera Mountains, is almost perfectly circular and over 600 miles in diameter rising to 20,000 feet above the plain, quite comparable to Mare Imbrium in diameter. The inner ring of Rook Mountains is some 400 miles in diameter. Material has been thrown beyond the Cordillera scarp for another 600 miles, constituting a unique blanket of complicated flow structures largely radiating outward from the center, obliterating most of the earlier surface features.

For both Mare Imbrium and the Orientale basin, gigantic impacting meteorites or comet nuclei, some tens of miles in diameter, have exploded craters also tens of miles deep and with diameters of hundreds of miles. The subsequent healing process of magmatic readjustment proceeded more completely for Mare Imbrium. Mare

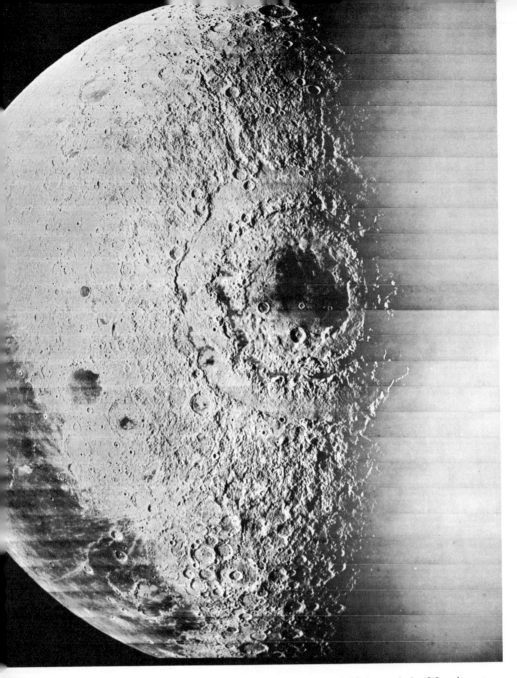

Fig. 115. The Great Orientale Basin centered 9° beyond the Moon's eastern limb 15° south of the equator, as viewed by U.S. Lunar Orbiter IV. Oceanus Procellarum appears in lower left. (Courtesy U.S. National Aeronautics and Space Administration.)

Orientale appears to be much more recent. Whether the temperature of the outer mantle had been reduced by that time, whether the underlying material was fundamentally cooler, or whether the somewhat smaller size of Mare Orientale accounts for its more perfect preservation, detailed geologic studies of the Moon will eventually clarify. The concentric ring structures of mountain scarps and collapsed surfaces have aroused a number of theoretical speculations. Can they result from some supersonic shock wave phenomenon of hyperballistic impact, or do they literally reflect stratification layers at various depths beneath the lunar surface?

In this concept that the maria were formed as the result of meteoritic collisions coupled with great lava flows, we need not postulate that the lava all came from the central crater at the point of impact. Such violent disturbances on the Moon, capable of producing great cracks at distances of hundreds of miles, would naturally provide the means of allowing any lava near the surface of the Moon to escape through innumerable fissures. It could thus fill up these cracks as well as penetrate the floors of older craters and generally patch up the low-level irregularities in the Moon's surface without many flows over great distances. Note that solid rock is denser than its melt, so that general subsidence of the solid surface would be expected under these circumstances.

Color photographs of the Moon with extremely accentuated color contrast by E. Whitaker show patchy irregular patterns over the maria basins. The edges of these patterns match the edges of very low-lying plateau areas, leaving little doubt that we are observing the records, of old lava flows. Whether these are dust or liquid lava flows can eventually be ascertained by direct study on the Moon.

The importance of *proving* that the maria are large lava flows is vital to understanding the Moon's origin and history. If the Moon had once been thoroughly heated, the iron should have melted with the silicate-type rocks and settled to the center, forming an iron core such as we believe exists on the Earth and for which we have clear evidence in the iron meteorites. The difficulty with this general viewpoint for the Moon is the fact that the Moon's mean density is only 3.33 times that of water. Pressure compression inside the Moon can hardly account for more than a 1-percent increase in density, which leaves the average density of the Moon much like that of ultrabasic rock found at moderate depths in the Earth. It is very difficult, then, to account for such a low density if the Moon contains as large a percentage of iron as does the Earth. The average stony meteorites, on the other hand, contain even a larger percentage of iron than does the Earth but have much the same density as the Moon. We must conclude either that the Moon, by some chance, contains a relatively smaller amount of iron than the Earth, or that the materials of the Moon generally have not been melted to separate the iron systematically. Other possibilities exist, it is true, but they become rather speculative in character.

If the Moon had been formed with as much average radioactivity as the Earth and the stony meteorites, it would have melted. In that case, as for the Earth, the radioactive materials, because of their chemical nature, would have been carried to the outer mantle with the lighter silicate materials. If much of the Moon has been melted, the maria should show relatively strong radioactivity like the upper mantle rocks of the Earth and not like the stony meteorites. The lunar space program now provides a specific answer to this vital problem. The U.S.S.R. spacecraft Luna 10 and 12, orbiting about the Moon, measured gamma rays near the Moon. Gamma rays are *very* high-energy light photons, more energetic than x-rays. They are produced by radioactive elements such as Uranium and by cosmic rays penetrating the surface and interacting with atomic nuclei. The Soviet scientists conclude that rocks like stony meteorites could not have produced the gamma rays observed from the lunar maria; rather they must have been produced by *basic* rocks like basalt or volcanic magmas from the outer mantle of the Earth.

The U.S.S.R. result was confirmed and amplified by the alpha-particle chemical experiment carried out on the Moon's surface by U.S. Surveyor V in August 1967. In this experiment the energizing alpha particles, or helium nuclei, were supplied by the man-made element, curium-242, radioactive with a half life of 163^d. When these alpha particles strike the nuclei of lunar atoms, some emit protons (hydrogen nuclei) of specific energies, and most scatter the alpha particles in patterns that identify the nuclei. Oxygen, 0.58 by number of atoms, aluminum 0.06, silicon 0.18, and magnesium were identified. Elements heavier than silicon accounted for 0.13 of the number of atoms, while carbon and the combined numbers of iron, nickel, and cobalt atoms constituted less than 0.03 each. Less than 0.005 of the atoms were heavier than nickel. The very "skin" of the top layer on Oceanus Procellarum was sampled, and it is definitely like basalt or basic lava. The maria surface is *not* like a stony meteorite. On the other hand, the Soviet scientists find tentatively that lunar highlands do *not* contain appreciable radio-active elements such as uranium, thorium, and potassium. Hence the highlands *may* be primitive material like stony meteorites.

The recent ground-based measures of the lunar shape are in mutual disagreement, as are various solutions for the irregularities of lunar gravity derived from the motions of lunar orbiting sub-satellites. Thus we cannot yet conclude with confidence how uniformly the Moon's material is distributed within its body. Furthermore the Moon shows no magnetic "bow wave" where it meets the solar wind. Were there much electrically conducting matter such as molten rock or solid iron within the Moon, the magnetic field lines of the solar wind should be distorted around the Moon. Furthermore, both U.S.S.R. and U.S. spacecraft can find no magnetic field on the Moon exceeding 0.001 that of the Earth, thus providing further evidence against molten circulating material or large volumes of magnetic iron within the Moon.

Thus we are left with a dilemma. The visual and chemical evidence indicates the presence of past lava flows on the Moon, while the interior seems not to be molten, nor near the melting point for rock. Must we conclude from all this evidence that the Moon never melted as a whole but that local melting occurred at or near the surface? The "geological" study of the Moon will, indeed, rival the complexity of such studies on the Earth. The terrestrial stratified layers of recurrent sedimentation will not be duplicated on the Moon, although overlaying lava flows may be frequent, especially on the Maria. Apparently water erosion, although possibly present, has not been a significant factor in molding the Moon's present surface, whether or not the Moon once had a significant atmosphere and water. Stratification on the Moon has certainly occurred, however, both in large scale by meteoritic impacts such as those producing Mare Imbrium and Mare Orientale and in smaller scale around *all* impact craters. Furthermore, the maria show clear external evidence of recurrent plutonic activity. Thus the astrogeologist can anticipate no end to rewarding exploration.

Tide-raising forces and the evidence presented in Chapter 7 for the lengthening of the Earth's day indicate that in the past the Moon must have been much closer to us. Sir George Darwin (1845–1912), son of the great naturalist, theorized that the Moon was once in contact with the Earth, the two bodies having a period of rotation of about 4 hours, and that subsequently they separated because of resonant tide-raising effects. Darwin's complete theory cannot be supported because viscosity in the Earth-Moon would have prevented their separation. Also we now have strong evidence that the Earth and probably the Moon both formed as cool bodies and, therefore, separately. There is, however, no clear indication concerning the actual distance from the Earth at which the Moon was formed. Probably it formed at much less than half its present distance and possibly quite close to the Earth. Once there was hope that a fossil tide might be preserved in the Moon but we have seen that the Moon is not rigid enough for this.

Because tide-raising effects vary as the inverse cube of the distance, the rate of the Moon's travel from the Earth was very much faster in the early stages than it is today. Indeed, there may have been a time when the tide-raising effects on the Earth were far greater than those that we now observe. Kuiper even suggests that these tides were so high—of the order of a mile—that they had profound effects on the topography of the ocean bottoms, assuming that the Earth indeed had water at this early stage in its development.

We shall later (Chapter 14) speculate a bit more on possible mechanisms of lunar formation and evolution. Direct exploration of the Moon is continually answering old questions and raising many new ones. Landings on the Moon have already penetrated some of the secrets of the Goddess of the Night. Her ancient skin carries a record that predates any now left on Earth and her story probably reaches back to the days when the Earth was new.

10

Jupiter, the Dominating Planet

From the barren, inflexible features of the Moon, we now turn our attention to the antithesis of everything lunar—to the colossus of planets, Jupiter, whose surface presents a turmoil of never-ceasing transmutations. We find this planet some 400 million miles away in space; its mass amounts to more than 300 Earths, while its volume exceeds that of our planet by more than a thousandfold. Through the telescope we can see Jupiter as a golden disk with dark and light bands roughly parallel to each other. Reddish or brown shades of color catch our eyes while we note the irregular cloudlike patches that break the uniformity of the bands (Fig. 116). The disk seems slightly elongated in the direction of the bands, and careful measures confirm our judgment; this diameter is greater than its normal by one part in fifteen.

Within an hour's observation the planet appears to have turned appreciably, as in Fig. 116; in only 9 hours 55 minutes it will have made a complete rotation. On a succeeding night we find that the bands and surface markings are much the same as before, but that

Fig. 116. Jupiter. Photographs by the 200-inch reflector showing the rotation in 50 minutes and the motion of a satellite's shadow on the disk. (Photograph by the Mount Wilson and Palomar Observatories.)

the details are slightly changed (compare the two photographs in Fig. 117). Within a few weeks the band structure is considerably transformed though the general character of the markings remains unaltered. Since the axis of rotation is perpendicular to the plane of the slowly changing bands, the bands must result from great "trade winds" or atmospheric currents parallel to the equator.

The rotation of Jupiter is indeed rapid. Near the equator there is a zone where the rotation appears faster than in higher latitudes. There the period of revolution is some 5 minutes shorter, only about 9 hours 50 minutes. Periods between 9 hours 51 minutes and 9 hours 53 minutes are rarely observed, and then not persistently. The equator turns with a velocity of about 25,000 miles per hour. The consequent centrifugal force of rotation is sufficient, even when acting against 2.6 times the surface gravity on Earth, to flatten the sphere by the appreciable amount that we have already noted. The amount of flattening, however, is not as great as would be expected were the interior of Jupiter similar to the interior of the Earth. Involved calculations show that the ratio of the density near the center of Jupiter to the density of the upper levels must be greater than for the Earth. Thus the rate of increase in density from the surface to the center is relatively more rapid for Jupiter.

This peculiar density distribution in Jupiter acquires a striking significance when we recall that the mean density of the entire planet is only 1.33 times that of water. The high concentration of material toward the center requires that the outer layers be much

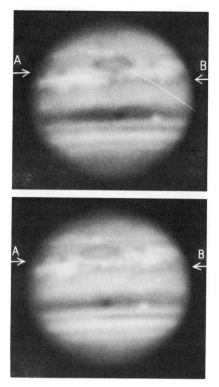

Fig. 117. Jupiter in 1928. The photographs, taken 49½ hours apart, show changes in the Great Red Spot and motions of clouds near it. Note spots indicated by arrows. (Photograph by E. C. Slipher, Lowell Observatory.)

less dense than water. Since there are few solids or liquids so rare, we must conclude that the outer layers of Jupiter are gaseous or are composed of exceedingly light materials.

The existence of a very deep atmosphere is evidenced by certain of the directly observed markings. Most startling of all the features on the Jovian surface has been the Great Red Spot, first noted in 1878. It appeared as a brick-red area, elongated some 30,000 miles (nearly 4 times the Earth's diameter!) in a direction parallel to the equator. It is still observable as an oval spot of varying color and character, but has never been as conspicuous as during the first few years after discovery. Figure 118 presents the aspect of Jupiter and the Red Spot as photographed in blue and in red light. In blue light the Spot is conspicuously dark against the planet's disk. In red light the Spot has almost disappeared. A white object would

Fig. 118. Jupiter photographed with the 200-inch reflector in blue light (*upper*) and red light (*lower*). Note that in red light the Great Red Spot has essentially disappeared. Ganymede and its shadow are also shown. (Photograph by the Mount Wilson and Palomar Observatories.)

show equally well in both colors, while a red one fails to reflect the blue light; hence it appears dark when photographed in blue light. The impartial "eye" of the photographic plate therefore verifies the color of the Great Red Spot. The general reddish tinge of other markings is also evidenced by Fig. 118.

The Great Red Spot does not rotate uniformly with the planet but drifts about considerably; the most assiduous student of Jupiter, B. M. Peek, finds that it has wandered as much as three revolutions from its average position on the planet calculated with a constant period. Such freedom of motion shows unquestionably that the Spot is a floating disturbance. Other such semipermanent markings are observed on Jupiter's surface, and all wander about to some extent.

We know directly three of the chemical constituents of Jupiter's atmosphere; they are the element hydrogen and the compounds ammonia and methane (or marsh gas). The determination of the chemical composition of a gas 400 million miles away is indeed a feat of scientific magic. It is done with a spectrograph, an instrument that separates light into its fundamental colors and photographs the entire sequence, from the ultraviolet through the blue, green, yellow, red, and infrared. The light to be analyzed first passes through the entrance slit of the spectrograph (Fig. 119). then through a lens to a prism, the heart of the instrument. The prism disperses the light into its constituent colors, forming a spectrum. The spectrum is identical in character to a rainbow except that the colors are much better separated. A second lens of the spectrograph serves to focus the spectrum on the photographic plate, where a permanent record is made. With a long exposure the photographic emulsion registers light much too faint to be seen with the naked eye.

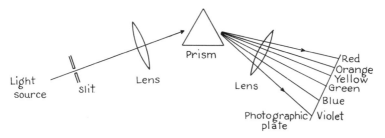

Fig. 119. Schematic drawing of a spectrograph. The optical functions of the lenses are not indicated.

The spectrograph solves a number of our difficult problems, one of the most important being the identification of chemicals in a gaseous mixture. The light that escapes from a luminous or absorbing gas reveals the character of the gas. Each atom or molecule is a vibrating pulsing entity; the vibrations are those of the elementary negatively charged particles, the electrons, which are held in miniature orbits or energy states about the heavier positively charged nuclei of the atoms. Two or more atoms in a molecule also rotate and vibrate around each other, while the electrons in the atoms spin about in their complex gyrations.

All of these motions and vibrations in atoms and molecules represent discrete amounts of energy, exceedingly minute but characteristic for each kind of atom or molecule. When an atom loses energy by a change in the rate of vibration, the energy is radiated into space as a *photon* of light (or *quantum*, the smallest unit of radiant energy), with a definite amount of energy, a certain color to the eye (if visible), and a certain wavelength of vibration. These waves of light belong to the same family as radio waves but are much shorter. The waves of red light are about 26 millionths of an inch long, while those of blue-green light have a length of only 19 millionths. Infrared or heat waves may be several times longer; ultraviolet light waves are shorter.

The compressed incandescent gas near the surface of the Sun sends out all colors and therefore all wavelengths of light. When such continuous light traverses a layer of cool gas, such as our atmosphere or the atmosphere of Jupiter, the vibrating atoms and molecules are activated, and steal from the beam precisely those wavelengths characteristic of their rates of vibration. When we analyze the light with a spectrograph we measure the wavelengths that are missing, and so identify the atomic or molecular thieves who are responsible for the loss. The missing wavelengths appear as dark *lines* in a spectrum.

In Jupiter's atmosphere the molecules of ammonia and methane are responsible for the losses shown by the spectra of Fig. 120. Methane predominates in the spectra of other giant planets. Other dark lines to be seen in these spectra have been produced by the gases of the Sun's outer layer and by the Earth's atmosphere. Theodore Dunham of the Mount Wilson Observatory was able to identify these two gases by separately compressing them in a 60-foot pipe. He discovered that the wavelengths lost from a light beam

Blue   Green   Yellow   Red   Infrared

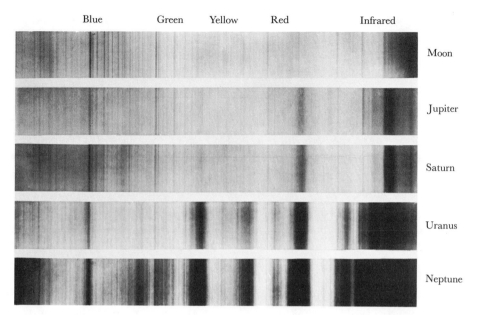

Moon

Jupiter

Saturn

Uranus

Neptune

Fig. 120. Spectrograms of the Moon and the giant planets. The great dark absorption bands, most conspicuous in Neptune's spectrum, are due to methane. (Spectra by V. M. Slipher, Lowell Observatory.)

reflected twice through the pipe agreed identically with the wavelengths absent in the spectrum of Jupiter. Some 30 feet of ammonia gas at standard atmospheric pressure are equivalent to the amount in Jupiter's atmosphere, to the depth that sunlight penetrates before it is reflected back to us. For methane, the corresponding amount is about 0.1 mile.

Now the mere presence of these particular gases tells us immediately that hydrogen is abundant in Jupiter's atmosphere. Ammonia (NH_3) is composed of one nitrogen atom to three of hydrogen, while methane (CH_4) contains one carbon atom to four of hydrogen. We may conclude that hydrogen is so abundant that it combines with all the carbon and nitrogen present. In 1963 H. Spinrad and L. M. Trafton discovered a special type of absorption band due to the hydrogen molecule. More than a 40-mile equivalent of hydrogen at sea-level pressure occurs above the clouds of Jupiter. Probably hydrogen has also combined with the available oxygen to form water (H_2O), but the water has frozen and sunk to the bottom of the atmosphere, out of our sight. If so, there may well be a thick layer of ice far beneath the clouds on Jupiter.

To learn more about the nature of Jupiter's atmosphere, W. A. Baum and C. D. Code observed with the 60-inch reflector of the Mount Wilson Observatory Jupiter's occultation of a fifth-magnitude star, Sigma Arietis, on November 20, 1952. From the rate at which the star's light was occulted they calculated the refraction of the light by the atmosphere of Jupiter and hence determined the rate at which atmospheric density fell off with increasing height. They found that the atmospheric density drops by a factor of 2 when the altitude increases by a distance of 2.8 to 5.5 miles. In an atmosphere of known temperature and gravity, this information leads to a knowledge of the mean molecular weight of the atmospheric gases. For example, molecular hydrogen has a total weight of 2, helium 4, carbon 12, methane 16, nitrogen 14, ammonia 17, oxygen 16, water 18, and so on. With a reasonably accurate knowledge of the temperature of Jupiter's atmosphere Baum and Code find that the mean molecular weight is in the range from 1.8 to 5. From spectroscopic evidence J. E. Beckman concludes that the mean molecular weight is 2.65, making hydrogen the major constituent of the atmosphere above the cloud tops. Helium then appears to constitute about half the remaining mass, much like its abundance in the Sun, while methane constitutes only a trace, a fraction of a percent by mass.

To understand how such a massive planet as Jupiter can have a mean density as low as 1.33 times the density of water, we would naturally expect to make it mostly of hydrogen because hydrogen is the lightest element. A schematic model of Jupiter's internal constitution was deduced in 1943 by R. Wildt, now of Yale University. In this model the outer layers, for the first 18 percent of the radius, consisted chiefly of compressed hydrogen; an ice layer filled the next 39 percent, and a metallic rocky core constituted the remaining 43 percent. The ice was compressed by the enormous pressures to a density of 1.5 times that of water, the hydrogen to 0.35, while the core was assumed to have a density of 6.0. In 1948, however, W. H. Ramsey in England calculated that Jupiter *could* be made entirely of hydrogen, which becomes a solid *metal* at a pressure of some 800,000 atmospheres. Only a hydrogen planet with a mass of more than 70 Earth masses can develop such a metallic-hydrogen core. A good model approximation to Jupiter contains about 80 percent of hydrogen and a few Earth masses of heavy earthy atoms, with the remainder helium. W. C. DeMarcus allows a central core of heavy materials extending to about 0.1 of the radius and possessing a central density of 30 times water. P. J. E. Peebles calculates models composed of hydrogen 0.80, helium 0.18, and earthy material 0.02 by mass. The fraction of helium he postulates is about half that of the Sun.

Jupiter, incidentally, is almost the largest planet possible. If it were much more massive it would shrink in size because of the formation of a dense core of degenerate matter, or, if more massive still, become a radiating star. If it were appreciably less massive, or made out of denser materials, it would again shrink in diameter. Planets somewhat like those of the Sun have been proved to exist about other stars. Peter van de Kamp showed in 1963 that the nearby star, Barnard's "runaway" star, has a companion planet about 1.6 times the mass of Jupiter. This planet may actually be smaller in diameter than Jupiter.

The composition of the great clouds is no longer a matter of pure speculation. The known abundance of ammonia and methane, combined with a knowledge of the temperature at Jupiter's surface, provides sufficient clues to reveal the secret. The surface temperature is about $-227°F$, approximately the temperature we should expect if the surface is heated solely by the Sun. At ordinary atmosphere pressure ammonia boils at $-28°F$ and freezes at $-108°F$, while methane boils at $-259°F$ and freezes at about $-300°F$. Hence ammonia on Jupiter is frozen, while the methane is gaseous. The clouds, consequently, must contain small ammonia crystals suspended in the atmosphere, as ice crystals are held in terrestrial clouds. The ammonia vapor that we observe has *sublimed*, that is, evaporated without melting from the solid crystals, otherwise no ammonia gas should be detected.

A. B. Binder and D. P. Cruikshank at the Kitt Peak National Observatory have looked for a brightening on Jupiter's surface immediately after a satellite's shadow had passed. The cooling might produce new reflecting crystals in the atmosphere. For Io they found a brightening of 0.09 magnitudes, and for Europa, 0.03 magnitudes, lasting for some 10–15 minutes. Ganymede gave no effect.

The strong coloring of some of the clouds has been remarked earlier in this chapter. The photographs of Fig. 118 portray distinct variations in shades and tints over the several belts and clouded areas. The equatorial band is dark in blue light, but much less striking in red. It is yellow in color. A careful comparison of these excellent photographs will disclose a wide range of hues in addition to the brick-red of the Great Red Spot. Stable atmospheric compounds cannot produce the colors seen on Jupiter. The colors must form, circulate, and disappear; otherwise the planet would assume a constant hue. Possibly colored metallic contaminations might be thrown up from below, say by volcanos, finally to settle back or be altered chemically by the atmosphere. R. Wildt has suggested sodium as a possible contaminant, and H. C. Urey has suggested that organic molecules color the clouds on the giant planets. Indeed C. Sagan and S. L. Miller have produced brightly colored organic molecules by spark discharges through a simulated Jupiter atmosphere in the laboratory. They suggest that lightning in such cloudy, turbulent, giant atmospheres produces chemically short-lived colored compounds.

The region of Jupiter most difficult to visualize is perhaps the upper transition layer between the clouded atmosphere and the solids beneath. The gravitational force on the upper gases produces a rapid increase of pressure at greater depths in the atmosphere, and introduces the difficult problem of calculating the level at which the gases and clouds will form a solid substratum. A gas, such as methane, may remain vaporous even after being compressed far above its usual solid density, if the temperature increases with the depth—as is quite likely on Jupiter. The observed temperature of the planet, in agreement with expectations, shows only that Jupiter does not radiate *much* heat; Öpik suggests that it radiates as much internal heat as it receives from the Sun. An inappreciable contraction of 1 millimeter per year could release gravitational energy for this purpose. Thus the lower levels may be considerably warmer than the upper.

The late B. Lyot of France earlier applied another powerful technique to the study of Jupiter and other celestial bodies; he measured the *polarization* of their light. A light photon or quantum has a plane of vibration. Special filters, such as Polaroid sunglasses, for example, have the property of passing light quanta vibrating in one plane (Fig. 121) and absorbing light vibrating in a plane at right angles. In almost all reflection of light by surfaces the degree of reflection depends upon the geometry of the reflection, the nature of the surface, and the original polarization of the light. Sunlight and light from most common sources are unpolarized; that is, the light quanta vibrate randomly in all planes perpendicular to the

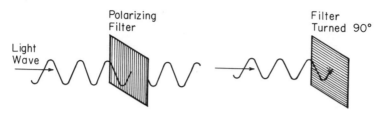

Fig. 121. Polarization of light. A light wave whose plane of vibration matches that of a polarizing filter can pass through, while if the filter is turned 90° the light is stopped.

axis of motion. Hence the polarization of scattered or reflected sunlight can frequently give us considerable information about the nature of the scattering or reflecting material. Thus sky light shows strong polarization, as does light reflected at grazing incidence from a surface, as sunlight may be reflected from a road.

A. Dollfus of France presents almost conclusive evidence from polarization measures that Jupiter's atmosphere contains a haze of fine particles, most concentrated in the equatorial bright zones, less over dark belts, and absent over the poles.

It is possible, too, that no distinct solid surface exists, that at some distance below the clouds the gases become very dense until a thick slushy layer (perhaps solid-ammonia particles) begins. The slush becomes heavier with depth until it is effectively solid. This occurs perhaps 50 miles below the tops of the clouds.

The continued existence of the Great Red Spot and other semi-permanent markings adds another complexity to the problem of whether Jupiter may have a solid surface. After the Red Spot became conspicuous it could be identified on drawings as early as 1831. The South Tropical Disturbance, less conspicuous than the Red Spot, moves somewhat irregularly around Jupiter's equator in about 2 years. It approaches the Red Spot with a speed of several miles per hour, hastens to catch up with the Spot, then seems to pull the Spot along for some distance. After the encounter the Spot drifts back while the Disturbance continues on its way. Such motions make it difficult to argue that a continuous volcanic action is responsible for the South Tropical Disturbance, or even for the Red Spot. On the other hand, it is difficult to postulate another type of source for such strange persistent clouds. Peek visualizes it as some strange type of floating solid, or at least a mass having a distinct solid nucleus.

Fig. 122. Rapid changes on Jupiter. These photographs were taken (*left to right*) on September 14, October 6, and November 30, 1928. Note the relative motion of the small white spot just below the equator. (Photographs by E. C. Slipher, Lowell Observatory.)

Relative motions of a similar type are apparent in the photographs of Fig. 122. A small white spot, just north of (below) Jupiter's equator, advances upon the dark spot slightly south and to the east of it (up and to the left). The gain is appreciable in the 3 weeks between the first and second photographs. Figure 117, earlier in this chapter, depicts relative motions of 8000 miles per day between the Great Red Spot and darker spots near it. The small interval between the exposures and the less conspicuous character of the small spots render the motions more difficult to detect in the reproductions.

Over periods of years Jupiter's surface features change completely except for the presently persisting Red Spot (Fig. 123) and the locations of the major belts.

Radio astronomy has produced some remarkable new knowledge about Jupiter and raised some interesting questions. In 1955 B. F. Burke and F. L. Franklin of the Carnegie Institution of Washington were studying radio stars with an antenna system tuned to a wavelength of 13.5 meters when they discovered that Jupiter was also a radio star producing noise at semiregular intervals. C. A. Shain of Australia then searched records of Jupiter noise back to 1951 and discovered that the noise could largely be associated with a certain area of Jupiter's surface (see Fig. 124). Even though the noise is not always radiated at each rotation of Jupiter, certain areas do produce the noise systematically and indicate a solid-body rotation for the sources, with a period of 9 hours 55 minutes 29 sec-

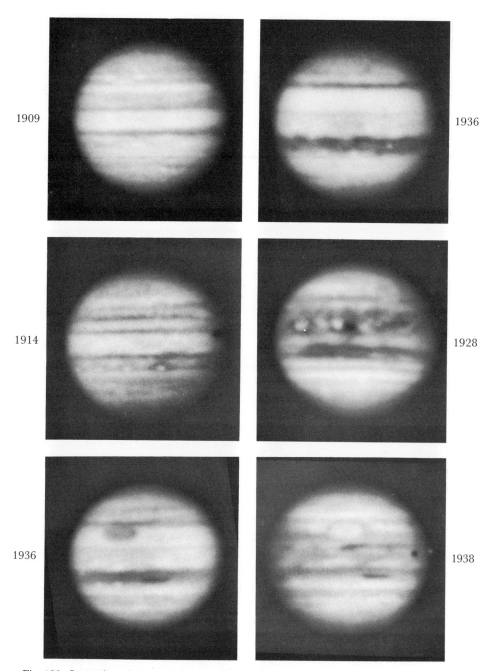

Fig. 123. Long-time changes on Jupiter. The photographs were taken in the years indicated. Shadows of satellites appear on the 1914 and 1938 photographs. (Photographs by E. C. Slipher, Lowell Observatory.)

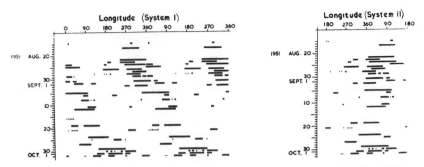

Fig. 124. Times of radio noise from Jupiter in 1951, as displayed by C. A. Shain. In the left-hand figure the times form a slanted structure because the observed surface period of rotation was adopted. The corrected period of rotation on the right brings the noise into synchronism with the solid (?) inner core.

onds, which appears to be quite constant. Hence there is clear-cut evidence that Jupiter does indeed rotate like a solid body and that the clouds we see at the surface present systematic atmospheric motions like our trade winds. No surface features are yet identified with these bursts of noise.

The great radio-noise bursts from Jupiter correspond in energy to a billion simultaneous lightning flashes on the Earth and seem to be of very short duration, a small fraction of a second. They are hardly ever observed beyond a wavelength of 20 meters, although the Earth's ionosphere begins to cut down seriously on their intensity at the longer wavelengths.

At microwave radio frequencies the temperature of Jupiter comes out very near to the expected value from infrared measurements, but, as the wavelength increases, the calculated effective temperature becomes much greater and indicates that Jupiter, like the Earth, has a magnetic field and a radiation belt in which very high-frequency radio noise is continuously generated. The magnetic field around Jupiter is of the order of ten times that of the Earth and the magnetic axis, like that of the Earth, is tilted at a considerable angle to the axis of rotation, perhaps 8 degrees. This, indeed, partially accounts for some of the semiperiodic character of the radio bursts at long wavelengths. No definitive answer is yet available as to the nature of the radio bursts but the existence of radiation belts is well established. It is not surprising that Jupiter, with its very rapid rate of rotation, should possess a strong magnetic field which produces on a much grander scale the same phenomenon that we see about the Earth. The longer wavelength radiation

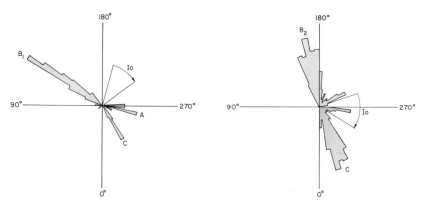

Fig. 125. Longitudes of Io (*left*, 195° to 235°; *right*, 250° to 300°) at which regions B_1, B_2, and C radiate 10-meter radio noise when they are on the central meridian as seen from Earth. The figure lies in the plane of the ecliptic with longitude in System III. (After C. N. Olsson and A. G. Smith.)

is highly polarized and is clearly a nonthermal type produced by high-energy electrons spiraling in a magnetic field. We observe the radiation when radiating beams from Jupiter's magnetosphere cross the Earth.

The mystery of Jupiter's radio noise was compounded in 1964 when E. R. Bigg discovered that radiation in the 10-meter range occurred only at specific longitudes of the inner Gallilean satellite Io. Figure 125 shows the amazing relation between the longitudes around Jupiter at which specific regions A, B_1, B_2, and C on Jupiter radiate when Io lies in two ranges of longitude. The radio-noise bursts seldom occur except when Io crosses the plane of Jupiter's magnetic field on only one side of the planet as seen from the Earth. The other satellites seem to be scarcely, if at all, involved in the phenomenon, which does not show in the centimeter band of wavelengths. We do not yet understand how Io can produce the effect.

Jupiter's Satellites

Jupiter's family of satellites is the most numerous in our system and could easily provide material for lifetimes of study. All the dynamic problems of the entire solar system are here reproduced on a miniature scale (Fig. 126). Of the 12 satellites so far discovered, the four largest outshine the next in size by a thousandfold, and the

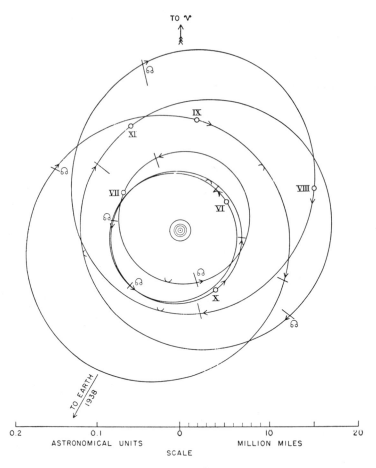

Fig. 126. Orbits of Jupiter's satellites. The four Galilean satellites and Satellite I occupy the small central orbits. Arrows indicate directions of motion. Arcs of small circles represent the intersections of the orbit planes with the ecliptic. The actual orbits do not intersect because of their various inclinations. (Diagram by S. B. Nicholson; courtesy of the Astronomical Society of the Pacific.)

faintest one by a factor of 100,000. Galileo observed the four satellites in January 1610, very shortly after he had constructed his first telescope; they are therefore called the Galilean satellites. They move in small orbits ranging from about one to five times the Moon's orbit in size. Because of the enormous mass of Jupiter, however, the periods of revolution are much shorter, from about 1 to 16 days. Of the fainter satellites, the fifth, discovered by Barnard in 1892, revolves with a period of half a day, while all the others have longer periods; the longest is more than 2 years.

The faintest satellites (IX, X, XI, and XII) can be photographed only by the larger telescopes, and were all discovered by the late Seth B. Nicholson (Fig. 127), of the Mount Wilson Observatory. Two appear as barely distinguishable white dots in Fig. 128. Since the satellites move in the sky and since long exposures are necessary, the telescope is made to follow the satellites and thus produces trailed star images.

The eighth, ninth, eleventh, and twelfth satellites are of particular interest because they are the outermost satellites (some 14,000,000 miles) and move in retrograde orbits, opposite in sense to the orbits of all other bodies (except comets) inside Jupiter's orbit. Jupiter itself rotates in the plane of its orbit, which is practically identical with the plane of the Earth's orbit. This contrariness on the part of the four satellites provides some material for speculation. Someone has suggested that they were once asteroids which, by chance, happened into precisely the proper positions near Jupiter to be captured by the gravitational attraction of this great planet. The theoretical stability of the present orbits is some measure of the chance of such a capture. It is true that the eighth satellite (retrograde) moves so far from Jupiter that at times its situation is

Fig. 127. Seth B. Nicholson (1891–1963), who discovered four of Jupiter's faintest satellites.

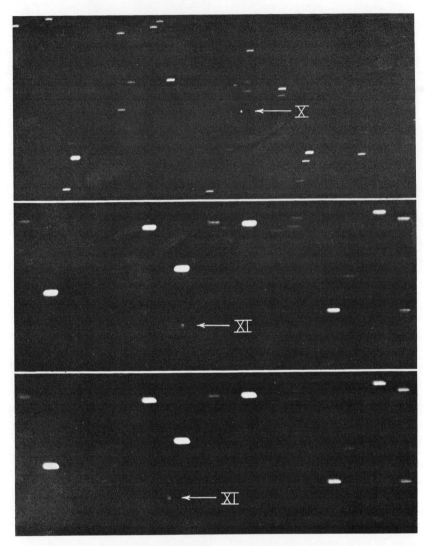

Fig. 128. Jupiter's satellites X and XI, photographed by S. B. Nicholson, their discoverer. (Photograph from the Mount Wilson and Palomar Observatories.)

very precarious because of the solar attraction, but it also appears that the retrograde motion of all four is a safeguard from loss to the Jovian system. There is still controversy as to asteroidal origin of the outer satellites.

The two inner Galilean satellites, Io and Europa, are similar to the Moon in size, mass, and density; they must, like the Moon, be composed largely of rock. The outer two, Ganymede and Callisto, are considerably larger than the Moon and somewhat more massive, with mean densities less than twice that of water. They may contain such materials as make up the outer layers of Jupiter, perhaps frozen ammonia and water, as well as rocky materials. During eclipse Ganymede cools very rapidly, like the Moon, suggesting a similar poorly conducting surface layer. Unfortunately the densities of the other Jovian satellites cannot be determined. It is thought-provoking, however, that the two inner Galilean satellites should be so much like the Moon while the two outer ones should contain such a large proportion of lighter materials. These facts must undoubtedly constitute an important clue to the formation, not only of the satellites, but of Jupiter itself.

All four of the Galilean satellites differ in one respect from the Moon: they are all much better reflectors of light, with albedos ranging from 0.54 for Io to 0.15 for Callisto. Their surfaces, therefore, must be quite different in character from that of the Moon, at least partially covered with some H_2O ice and also other frozen gases that would have long since escaped from the top surface of the Moon because of its nearness to the Sun. Little is known about the fainter Jovian satellites except that they are very small, the outermost ones being only a few miles in diameter.

The Galilean satellites have made possible one major contribution to physical knowledge. When Olaus Roemer (a Danish astronomer, 1644–1710) observed the eclipses of these satellites by Jupiter, he discovered that the time intervals between the eclipses were greater when the Earth was receding from Jupiter than when it was approaching. In 1675 he came to the conclusion that the apparent variations in the periods were caused by the *finite* velocity of light; previously light had been suspected of moving instantaneously. When the Earth is receding from Jupiter the light must travel a successively longer path between eclipses, while in approach the path is successively shortened. Roemer invented several of the most important instruments of positional astronomy. His great contributions, unfortunately, were little appreciated during his lifetime and only his proof of the finite velocity of light is much remembered today.

11

The Other Giants—Saturn, Uranus, and Neptune

Saturn

Among the innumerable celestial objects that may be seen through a telescope, the most beautiful of all is perhaps the planet Saturn. When viewed in the evening twilight while the sky is still bright, the yellow gold ball and its unbelievable rings shimmer in a brilliant blue medium, more like a rare work of art than a natural phenomenon. Lightly shaded surface bands, more uniform than those of Jupiter, parallel the great rings; only occasionally can one distinguish detailed markings that will reveal the rapid turning of the great globe. The central brilliance fades away toward the hazy limb of the planet's disk, and the rings at their borders appear to dissolve into the sky.

Where Saturn's rings cross the disk a hazy dark band outlines their innermost edge (Fig. 129). This "crape ring" is most readily discerned by its faint shadow on the planet's disk. The outer rings also cast shadows on Saturn, which in turn completely eclipses

Fig. 129. Saturn's rings disappear when seen edge-on. At left the rings are near their maximum opening in 1915. At right, the rings are edge-on, in 1921. See Fig. 121. (Photographs by E. C. Slipher, Lowell Observatory.)

large sections of the rings. The polar regions of the planet, perpendicular to the plane of the rings, are darker than the other edges of the disk, and, when seen under good observing conditions, present a slightly greenish appearance. Three major bright divisions of the rings are easily detected, the brilliant middle ring (*B*), the fainter outer ring (*A*), and the barely luminous crape or inner ring. The two outer rings are broken by narrow dark gaps (Fig. 130), similar to Cassini's division, which separates rings *A* and *B* (named for G. D. Cassini, 1625–1712, the first director of the Paris Observatory); these markings are detectable only under ideal observing conditions. Observers generally fail to distinguish the slightest irregularity or discontinuity around the ring surfaces except for the dark divisions concentric with the planet.

Saturn's rings lie precisely in the plane of the planet's equator, which is inclined some 28° to the plane of the Earth's orbit. Since the plane of the rings remains fixed as Saturn moves around the Sun, during one revolution we can see the rings from above (north), from

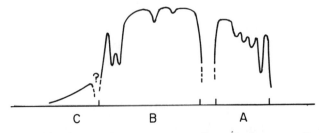

Fig. 130. The distribution of brightness across Saturn's rings, according to measures by A. Dollfus.

below (south), and twice edge-on (Fig. 131). When the rings are tipped at the greatest angle for terrestrial observation they reflect nearly twice as much sunlight as Saturn itself, but when the rings are edge-on they practically disappear for a short interval (Fig. 129). The thickness of the rings is therefore exceedingly small.

The bright ring B is some 145,000 miles in outer diameter and about 16,000 miles in breadth, while the outer ring A extends to a diameter of at least 170,000 miles and may be surrounded by very faint ring material to a far greater distance. Figure 130 gives a rather precise indication of the material content in the rings as measured by the surface brightness outward from about 7000 miles above Saturn's equator.

We know that these rings are composed of individual fragments of matter, each moving in its own orbit about Saturn according to Newton's law of gravitation. The irrefutable demonstration of this fact is an excellent example of scientific progress by the alliance of a physical theory with a special observing technique. The instrument used is again the spectrograph; the physical theory includes the laws of physical optics and of gravitation; the demonstration is the proof that any zone in the rings moves at precisely the speed that

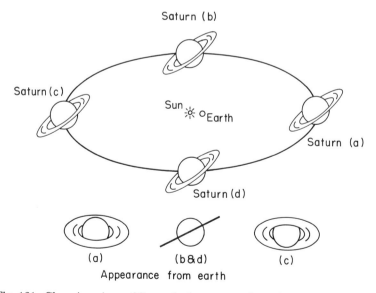

Fig. 131. Changing views of Saturn's rings as seen from the Earth. Positions (*a*) and (*b*) correspond respectively to the left-hand and right-hand photographs in Fig. 129.

would be expected were it composed of particles moving in circular orbits.

We have seen in the preceding chapter how the spectrograph can be used for identifying a gas by the spectral lines, or wavelengths, that are subtracted from the light when it passes through the gas. If the gas is approaching us, or we are approaching the gas, more of the waves enter the slit of our spectrograph in a given time interval than would do so if we were relatively at rest. The waves appear to be closer together because of the relative approach, with the result that all the wavelengths are measured shorter than before. Since wavelength decreases toward the violet end of the spectrum, all the dark lines of missing wavelengths are shifted toward the violet. The amount of the shift is proportional to the velocity of approach. When we are receding from the source of light the waves are apparently spaced farther apart, their lengths are greater, and the dark lines are displaced toward the red end of the spectrum. This phenomenon is known as the Doppler-Fizeau effect. In sound the analogous Doppler effect is exhibited by the drop in pitch of the bell of a train as it passes.

Measures of the wavelengths of the lines in the Sun's spectrum can be made on spectrograms of the Sun, and also when the sunlight has been reflected from a planet, satellite, or other object in the solar system. The shifts in wavelength are then a measure of the sum of the object's velocities with respect to the Sun and with respect to the Earth. The reflection naturally introduces the effect of the object's motion relative to the Sun, a shift in wavelength that is already present when the light reaches the object, and adds to it the effect of the motion with respect to the Earth.

When the reflected sunlight from Saturn's rings is focused on the slit of the spectrograph, the motions in the line of sight can be measured accurately in miles per second. The tilts of the lines in the spectra of Fig. 132 display the shifts in the wavelengths to demonstrate the rotation of Jupiter and Saturn. Note the reverse tilt of the lines in the rings of Saturn as compared with the disk. At every point along the rings the measured velocity agrees exactly with that of a corresponding satellite if it were moving in a circular orbit, more slowly with increasing distance from the center. Kepler's laws of motion are obeyed precisely; the inner sections of the rings rotate more rapidly than the outer sections. Were the rings solid, the outer edges would move more rapidly than the inner. This demonstration of the discontinuous structure of Saturn's rings was

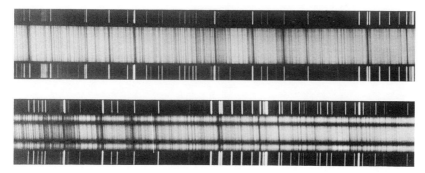

Fig. 132. Spectrograms of Jupiter and Saturn with its rings. Note the tilt in the lines of Jupiter's spectrum (*upper*) with respect to the comparison bright lines above and below, introduced by a spark. The tilt indicates differential radial velocity across the disk of Jupiter and hence its rapid rotation. Compare the spectrum of Saturn and its rings (*lower*) with Keeler's diagram, Fig. 133. (Spectrograms by V. M. Slipher, Lowell Observatory.)

performed in 1895 by J. E. Keeler (1857–1900) at the Allegheny Observatory. Keeler's diagram of the shifts of the spectral lines is shown in Fig. 133.

The exact nature and structure of Saturn's rings have constituted a major problem for more than 300 years. On rare occasions the rings will cross the line of sight to a star, occasionally the satellites pass behind the rings, and rarely we can see some of the satellites when shadowed by the rings. Stars disappear when obscured by the densest part of ring *B,* flash up to nearly normal brightness at Cassini's division, and show irregular fluctuations in ring *A,* much as might be expected from the brightness of the rings (Fig. 130). A small amount of light diffuses through even the densest part of ring *B* to give a very faint illumination of a satellite in the darkest shadow. F. A. Franklin finds that the rings are slightly redder than sunlight and that the thickest part of ring *B* has a high albedo of 0.70 in visual light and 0.57 in blue light. This observation is consistent with the rings being covered with ice or frost. Spectroscopic observations do not yet tell us the composition of the material.

The rings brighten up remarkably when observed almost at precise opposition from the Sun. Franklin finds that in the last degree there is an increase of more than 30 percent in the brightness of the rings. In 1887 H. von Seeliger (1849–1924) suggested that the many small particles in the rings shadow each other so that only at opposition can we see the surface of the particles directly without appreciable effects of shadowing. The phenomenon is similar to the

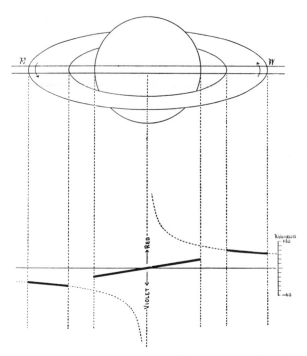

Fig. 133. Rotation of Saturn and its rings. Diagram to explain the spectrum to match Fig. 132. Note the tilt of the lines, showing that the outer edges of the rings move more slowly than the inner. The scale on the lower right represents radial velocity in kilometers per second. (J. E. Keeler drawing from the Yerkes Observatory.)

heiligenschein, the halo that appears about one's head in long shadows on dewy grass in the early morning (see Fig. 134). In matted vegetation, looking near the line of sight from the Sun one can see the light reflected directly from all surfaces. But at appreciable angles to the line of sight, the light entering the vegetation has to find a new route out to the eye and consequently is dimmed. To an observer in a high-altitude aircraft the shadow of an aircraft on the ground appears as a bright spot; over water it turns into a dark circle.

F. A. Franklin and A. F. Cook have recently based a new theory on improved photometry of Saturn's rings. They deduce that the rings consist of rather small particles, like sand, covered with a rough surface such as one might expect from frost crystals with dimensions of the order of $\frac{1}{300}$ inch. They conclude that the rings are very thin, perhaps a few yards to a few miles in thickness.

Fig. 134. The *heiligenschein* effect. Photographed by William Sinton from the dome of the 200-inch reflector. Look at the shadow of the dome and note how the brightness of the forest drops rapidly from the edge of the shadow.

It is doubtful that the specific material in the rings remains permanently in the structure. It is quite possible that the material of the rings is in a constant flux, ice and other materials being sublimed and replaced by new material captured from the interplanetary medium while the old is lost by effects of collisions, turbulence, gas pressure, and perturbations. Note the similarity of the problems of Saturn's rings to those of the lunar surface (Chapter 9), both optically and structurally, even though one consists of free particles in space and the other of particles attached to a surface.

Saturn is a unique member of the solar system, not only because of its rings, but also because its average density is less than that of water (0.7)! For Saturn, the problems of a deep atmosphere are even more complicated than they were for Jupiter. In spite of the low average density, the distribution of matter in Saturn is much like that in Jupiter—highly concentrated toward the center. The period of rotation, only 10¼ hours, and the polar flattening of 10 percent provide a measure of the concentration. In calculating a model for the interior of Saturn, W. C. DeMarcus finds that the hydrogen content (greater than 63 percent) is a little less than for Jupiter, in spite of Saturn's extremely low mean density. The smaller mass, about one-third of Jupiter's, does not produce such an enormous compression of hydrogen; consequently a somewhat larger core of earthy materials must be added to maintain even such a low value of the mean density. The central pressure, however, is still more than 50 million atmospheres.

The spectrograph reveals that Saturn's atmosphere contains more methane and much less ammonia than Jupiter's (note Fig. 120). Since Saturn ($-290°$F) is much colder than Jupiter, we may well conclude that more of the ammonia has frozen out of its atmosphere and that the reflected sunlight penetrates a thicker layer of methane. The lower temperature may also explain the more sluggish changes in the cloud formations and less complex structural detail. G. Münch and H. Spinrad at the Mount Wilson and Palomar Observatories have recently found direct spectrographic evidence for large quantities of molecular hydrogen in Saturn's atmosphere and conclude that the atmosphere consists primarily of hydrogen and helium.

Large-scale disturbances are occasionally manifest in Saturn's atmosphere. The great white spot of 1933 is conspicuous in the first photograph of Fig. 135; a year later, in the second photograph, the spot is missing and has been replaced by a white equatorial band. During the interval the northern (lower) hemisphere has completely changed its superficial appearance. The similarities in the atmospheric phenomena of Saturn and Jupiter are more striking than the differences, the latter being more in degree than in kind. This is true also of the colorings of the two planets, with the exception of an occasional greenish tint in Saturn's polar areas.

The photographs of Fig. 136 have registered Saturn's appearance in various colors of the spectrum. The white rings provide a ready

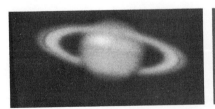

Fig. 135. (*Left*) The great white spot on Saturn in August 1933. (*Right*) The white equatorial band in September 1934. (Photographs by E. C. Slipher, Lowell Observatory.)

Fig. 136. Saturn in different colors: (*left*) red; (*middle*) blue; (*right*) yellow-green. Taken in 1940 with various filters and color-sensitive emulsions. (Photographs by E. C. Slipher, Lowell Observatory.)

standard for judging color shadings on the disk. Since the rings match the brilliancy of the equatorial belt in all three photographs, the belt must be nearly white. The next band in the southern (upper) hemisphere is particularly dark in the blue but shows brilliantly in the red and fairly strongly in the yellow-green. It is therefore orange-yellow, somewhat more reddish than the area above it. A greenish cast in the polar region is suggested by the comparative polar flattening in the three colors. From year to year the shadings of color over Saturn's disk are observed to vary greatly. Observations of the markings on Saturn made by many observers in many years show that the clouds on Saturn, like those on Jupiter, have a complex rotational pattern with a shorter period at the equator than at higher latitudes. Some 20 distinct latitudinal currents have been identified. The range in period is typically (e.g., as measured by L. J. Robinson) from 10 hours 13.6 minutes at the equator to 10 hours 40.1 minutes at latitude 60°. We still do not know whether the interior of Saturn is solid; no conspicuous radio static has been detected, as from Jupiter. The thermal radio radiation at 3.2-millimeter wavelength, however, indicates a low temperature, near the value obtained by infrared measurements, while the radio temperature increases with wavelength to room temperature in the 21-centimeter band.

On December 15, 1966, when the ring system of Saturn was edge-on with respect to the Earth and extremely faint, A. Dollfus was able to photograph an object of about magnitude 14 at a distance of some 98,000 miles from the center of Saturn. Other photographs almost conclusively confirm the reality of this tenth and innermost satellite of saturn, which Dollfus has named Janus. It moves in a nearly perfect circle in the ring plane with a period of 17.98 hours. The unexplained dip in the light curve of the rings (Fig. 130) lies at the position of two-thirds Janus' period, but a now-expected dip at one-half the period has not yet been observed. Janus is perhaps 190 miles in diameter. Awaiting complete confirmation of Janus' reality by observations when the rings are again edge-on, we shall maintain the old numbering of Saturn's satellites in other parts of this book.

Titan, the sixth satellite of Saturn's nine, is unique in the solar system. It has an atmosphere! In the winter of 1943–44, G. P. Kuiper, at the McDonald Observatory in Texas, photographed methane bands in the spectrum of Titan (Fig. 137) and found that its atmosphere contains about half as much methane as the atmospheres of Jupiter and Saturn. Since Titan is slightly smaller than Ganymede in Jupiter's system, with a diameter of 2990 miles com-

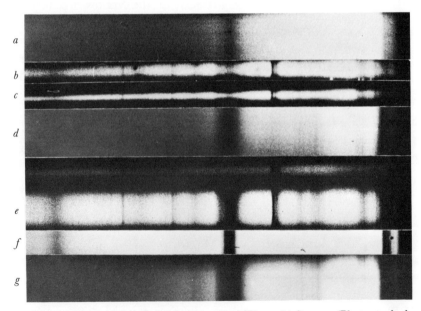

Fig. 137. Spectra: (*a, d, f, g*) ammonia; (*b, c*) Titan; (*e*) Saturn. (Photographs by G. P. Kuiper, from Yerkes Observatory.)

pared to 3120 miles, we may ask why, if Titan has an atmosphere, Ganymede does not? The velocity of escape does not favor Titan, 1.73 miles per second compared with 1.80 miles per second for Ganymede and 1.5 miles per second for the Moon. The fundamental difference lies in the lower temperature of Titan, some twice as far from the Sun as Ganymede. Allowing for the velocity of escape and the temperature as affected by albedo, Kuiper finds that Titan has a somewhat poorer ability to hold an atmosphere than Mars but a greater ability than Ganymede and that the dividing line occurs between Titan and Ganymede. He points out that if Titan's temperature were raised to $-100°F$ it would lose its methane atmosphere. Thus Titan provides an interesting, additional clue that may be important in finally ascertaining just how the satellites and planets came into being.

Japetus, the eighth satellite, much smaller than the Moon in diameter, is interesting because of its peculiar surface. It, like several of the other satellites, rotates with the same face toward Saturn, but, during a rotation of Japetus, its brightness varies by a factor of *six times*. Hence one side reflects light six times as well as the other; the surface structures of the two sides must be remarkably dissimilar. We might speculate that Japetus suffered disfiguration by a collision with some wandering member of the solar system, perhaps was partially discolored by gaseous outbursts from Saturn during the early stages of evolution, or possibly has a strikingly inhomogeneous composition.

The ninth and most remote satellite, Phoebe, revolves in a retrograde orbit 8 million miles in radius. When discovered by W. H. Pickering in 1898, it represented the only case of satellite motion retrograde with respect to its primary, but its claim to fame was later shared by other retrograde satellites. In the Saturn system the satellites are fairly similar in size and brightness; the faintest is Phoebe, which is some fiftyfold brighter than the faintest in Jupiter's family. Since Jupiter possesses so many satellites, it is rather surprising that more faint members have not been discovered in Saturn's system.

Five of Saturn's satellites, for which mass and density can be determined, have relatively low densities, all less than 2.4 times the density of water. They must all contain large quantities of ice, frozen ammonia, and other light materials mixed with rocky or earthy material.

Uranus and Neptune

These two planets are practically identical twins, giants in the outer regions of the solar system. Their diameters are about four times that of the Earth; Uranus is the larger by several hundred miles, although the measures are somewhat uncertain because of the hazy edges of the disks. Neptune, however, is the more massive, comprising 17.4 Earth masses while Uranus has 14.5.

Although surface markings are difficult to observe on Neptune (Fig. 138), perhaps because of the great distance, and only faint belts have been seen on Uranus, the planets are certainly enveloped with atmospheres resembling those of Jupiter and Saturn. The albedos are high and the spectra show methane absorptions similar to those for Jupiter and Saturn, but much intensified. Observe the sequence of spectra in Fig. 120. The absorptions of yellow and red light by methane vapor are so enormous for Uranus and Neptune that the planets appear greenish in color when observed directly; the color is more pronounced for Neptune. The spectrograms evidence no certain trace of ammonia, but hydrogen is present.

A lack of gaseous ammonia and an abundance of methane are readily explained as resulting from the vast distances of these planets and the corresponding diminution in the amount of solar heat received at their surfaces. The sunlight is indeed so weak for Uranus that its surface temperature is less than $-300°$F, while Neptune is probably more than 20 degrees colder. The temperature conditions on these outer planets are so extreme that laboratory techniques can duplicate them only on a very small scale.

The increasing strength of the methane absorptions and the weak-

Fig. 138. Surface markings on Neptune, compiled by A. Dollfus from drawings by three observers in 1948. No equatorial bands are detected.

ening of the ammonia absorptions as we progress from Jupiter to Neptune (again Fig. 120) certainly arise from the decrease in temperature. The vapor pressure of ammonia decreases very rapidly with temperature; hence precious little ammonia gas can remain in Neptune's atmosphere. Furthermore, with a decreased temperature, more of the ammonia crystals must settle out of the gaseous hydrogen strata, thereby reducing the number of clouds and bands, so conspicuous on Jupiter. Similarly, the interiors of the more distant and somewhat smaller planets are probably cooler. Thus the effects of internal eruptions or disturbances, such as Jupiter's Red Spot, are progressively less conspicuous on Saturn, Uranus, and Neptune. Variations in surface color, too, are less prevalent, none being observable on Neptune. The increased methane absorption with lowered temperature may result from greater penetration of sunlight into atmospheres less filled with clouds. The reflected light passes through thicker layers of methane.

The rotation periods of the twin giants must be determined by indirect methods because of the difficulty in observing surface markings on Uranus and the absence of markings on Neptune. In 1912 Percival Lowell and V. M. Slipher first utilized the spectrograph to measure the speed of rotation for Uranus. They found that the spectral lines at the edges of the planet's disk were displaced by an amount corresponding to a speed of about 10.5 miles per second. From the known circumference they deduced that Uranus rotates in 10¾ hours. Three years later, Leon Campbell (1881–1951) of the Harvard Observatory observed regular fluctuations in the brightness of Uranus and confirmed the spectrographic period. The best modern value is 10 hours 49 minutes.

Neptune's great distance reduces the size and brightness of its apparent disk to such an extent that the spectrograms are very difficult to obtain. At the Lick Observatory, J. H. Moore (1878–1949) and D. H. Menzel obtained the value of 15.8 hours for the period of rotation, just twice the period of small variations observed in the light. Neptune apparently had surface irregularities on opposite sides when the light fluctuations were measured.

Both Uranus and Neptune therefore rotate very rapidly. They are also flattened at the poles. Calculations from these data show that Uranus and Neptune are centrally condensed, like Jupiter and Saturn, and confirm the general opinion that the four planets are fundamentally similar. The sequence in densities—Jupiter

Fig. 139. Uranus and three of its satellites. (Courtesy of the Lick Observatory.)

1.33, Saturn 0.68, Uranus 1.60, and Neptune 2.25—suggests superficially that there is a discontinuity in structure as we move through the sequence. This does not follow, however, when one considers the effect of mass in compressing hydrogen. As we go from Jupiter to Neptune, the percentage of hydrogen decreases in each step. The extremely low density of Saturn arises simply from the fact that its mass is too small to compress the hydrogen adequately. Uranus and Neptune are more than ten times as dense as one would expect were they made entirely of hydrogen. Harrison Brown has suggested that they are composed mainly of ice and of solid ammonia and methane. It is not yet entirely clear how much earthy material is contained in these planets, but it may well exceed the amount in Jupiter and Saturn. Various investigators place the free hydrogen content of Uranus at about 10 percent and of Neptune at considerably less.

Uranus is anomalous in one respect. The plane of rotation of the planet, which is also the plane of revolution for its five satellites, is tipped nearly at right angles to their plane of revolution about the Sun. In fact the plane is tipped slightly more than at right angles (98°), so that all the motions are technically retrograde. The five satellites are very faint (Fig. 139), observable only in large telescopes, and all move in the direction of Uranus' rotation. Their diameters cannot be measured directly but are estimated to be of the order of a few hundred miles. The innermost, Miranda, was discovered by Kuiper (Fig. 140) in 1948.

Kuiper also discovered Neptune's smaller satellite, Nereid, in 1949. The larger, Triton, weighs about twice as much as the Moon, according to D. Brouwer and G. Clemence. Its diameter is too small to be measured but its brightness suggests that it is perhaps larger than the Moon. The two satellites show a remarkable disparity in their orbits. Whereas Triton moves in a retrograde motion about Neptune at an angle of some 40° to the plane of the planet's motion at a distance of some 220,000 miles, Nereid moves in the direct fashion at a distance of some 3,500,000 miles. Thus,

Fig. 140. Gerald P. Kuiper, who discovered the satellites Miranda of Uranus and Nereid of Neptune.

Triton is the only inner satellite in the solar system with a retrograde orbit.

We see that the giant planets are really very much alike, the major superficial differences being produced by the varying temperatures that arise from their positions in space. They all rotate rapidly, have huge atmospheres of methane and probably ammonia, and contain the light gases, helium and hydrogen. In all these characteristics they differ from the terrestrial planets, Mercury, Venus, Earth, Mars, and Pluto. The differences are so striking in every detail that it seems incongruous to associate these two groups of planets in the same system. Some major evolutionary factor was clearly responsible for these differences.

The giant planets, in spite of, or because of, their huge dimensions, provide no possible abode conducive to life of any kind now known. We must study the terrestrial planets if we hope to demonstrate universality for the ethereal phenomenon of life.

12

The Terrestrial Planets— Pluto, Mercury, and Venus

Pluto

The most distant planet as yet discovered belongs to a species entirely different from that of the other planets in these outer regions. Pluto appears to be a dwarf interloper among the giants. Our knowledge of Pluto is meager; besides the orbit and consequent distance we know Pluto's brightness and color, but not its mass (see Chapters 3 and 4). G. P. Kuiper has determined its apparent diameter to be in the range of 0.2–0.3 seconds of arc, corresponding to about 3600 miles. If we tentatively assume that the presently calculated mass, 0.8 that of the Earth, is approximately correct, then the mean density comes out greater than the density of gold!

Since metals and substances denser than iron appear to be rare in the stars as well as on the Earth, it seems utterly impossible that Pluto has a density much greater than that of iron, or 7.8 times water. It is far too small to develop a high-density core of *degenerate* matter, like the White Dwarf Stars. Either the determination of mass or that of diameter must be seriously in error.

D. Alter has suggested an alternative explanation, that Pluto's true diameter is greater than its apparent diameter because the planet has a relatively smooth surface so that the sunlight is reflected from a small area near its apparent center. Polished spheres or oval surfaces, for example, shine with concentrated high lights when illuminated with a point source. This solution, however, still does not solve the problem of Pluto's relatively low brightness. With a reflecting power equal to the very low value presented by the Moon, 0.07, Pluto should then be about twice as bright as it is observed to be. Furthermore, we should expect its surface to be rough and pitted because of influx from cometary debris.

The problem of Pluto's diameter now appears to be solved because of an observational program set in motion by I. Halliday of Canada. He predicted a possible occultation of a faint star by Pluto on April 28, 1965, and enlisted many observers in watching for the event. Since no one could observe a diminution in the star's brightness as close as 0.125 seconds of arc from the center of Pluto, its diameter must not exceed 4200 miles, supporting Kuiper's measure. The mass determination must be seriously in error, and the prediction of Pluto's existence was purely a matter of chance (see Chapter 3).

Pluto, therefore, is probably an arid, frigid, and rough little world, of diameter a little less than half that of the Earth and an albedo of some 0.15, twice that of the Moon. It is certainly an inhospitable planet for human habitation; its agonizingly cold night would last for 76.6 hours, followed by an equal period of daylight in which the Sun would shine with about $\frac{1}{1600}$ of its brilliancy as seen from the Earth. Its period of rotation has been measured only by its periodic variation in brightness.

It has even been suggested that Pluto is not truly a planet at all, but is a lost satellite from Neptune. This question cannot be answered until we know more about the mechanisms whereby satellites develop about planets. At least we need not worry about Pluto's destruction by collision with Neptune, even though their orbits overlap. E. Öpik, C. J. Cohen, and E. C. Hubbard have shown that Pluto and Neptune are held gravitationally in a repeating orbital cycle of nearly 20,000 years, so that they can never collide. If this relation holds rigorously, it would suggest that Pluto may not have once been a satellite of Neptune.

Mercury

Mercury is the fourth brightest planet, at its best nearly equaling Sirius in brilliancy and being exceeded only by Venus, Mars, and Jupiter. Nevertheless, Mercury is a very difficult object to observe because of its small orbit and concomitant proximity to the Sun; the greatest possible *elongation* (apparent angle from the Sun; see Appendix 2) is 28°. At this most favorable position the phase corresponds to the quarter moon; the full phase occurs at superior conjunction when Mercury lies beyond the Sun, nearly in line with it. After sunset or before sunrise Mercury is always very low in the sky, a situation that limits night observations to a short interval. In addition, the turbulence of our atmosphere at low altitudes produces poor "seeing." Hence Mercury, to a great extent, is observed in full daylight, scattered sunlight being eliminated as much as possible by suitable screens.

Because of these various difficulties only the most expert observers have been able to detect surface markings on Mercury. G. V. Schiaparelli (Italy, 1835–1910) and E. E. Barnard (U.S.A., 1857–1923), two great observers, each sketched vague surface detail, not in excellent agreement. Since the lunar photographs have shown that certain markings on the Moon, particularly the rays and to some extent the maria, are more conspicuous when the Moon is full, it is of considerable interest that Barnard described the markings on Mercury as generally similar to the maria on the Moon, and that Schiaparelli obtained his best results when the planet was at its full phase, close to the Sun. Mercury has been extensively observed in France, first by E. M. Antoniadi at Meudon, and more recently by A. Dollfus at the Pic du Midi.

Visual observers long ago agreed that Mercury keeps the same face towards the Sun and thus rotates on its axis in its period of revolution, 88 days. Radar observations from the great 1000-foot dish at Arecibo, Puerto Rico, however, have dispelled this long-standing illusion. Radar waves, scattering off a rough sphere like Mercury, can measure the range to a ring (Fig. 141) on the surface by control of the time lag from pulse transmission to pulse reception (10^{-6} second equals 492 feet). The diameter of a planet or satellite

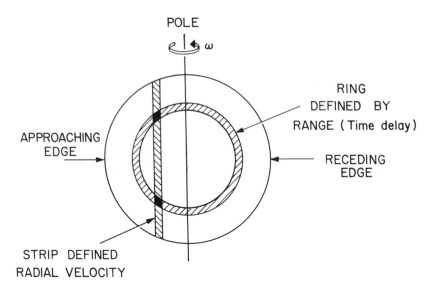

Fig. 141. A spherical planet illuminated by radar. The radar beam, broader than the apparent planetary diameter, is limited by time delay to a circle on the apparent disk and by frequency to a strip parallel to the rotation axis of the planet.

can be measured quite precisely by the length of the pulse echo corrected for the inherent length of the transmitted pulse. In addition the frequency or wavelength of the returning echo will be changed by the relative velocity between the source and target, the Doppler-Fizeau effect, described on p. 182. Thus, a rotating planet presents a higher returning frequency on its approaching side than on the receding side. A chosen frequency band can then be selected to measure the echo from a strip parallel to the rotation axis on the apparent planetary disk (Fig. 141). By choices of both time delay and received frequency, the combined echo from two areas (black in Fig. 141) common to the ring and strip can be measured. If the radar dish is large enough at a given frequency or, if not, by interference between two receiving systems, the beam can be narrowed enough to distinguish between the two dark areas. Thus radar maps of the Moon now rival the best Earth-based photographs. Soon excellent radar maps of Mercury, Venus, and Mars will be made. Diameters for these planets are now much more accurately determined by radar than by optical means from the ground.

For Mercury the radar-Doppler measures showed that the sidereal period of rotation is only 59 days direct, not 88 days as long believed. Immediately it became apparent that the period was nearly two-thirds the period of revolution and that the angular velocity about the Sun at perihelion would be very close to the average angular rotation. Since tidal forces vary as the inverse cube of the distance, the solar gravitational control on a slightly elongated Mercury body would mostly take place near perihelion and could stabilize the rotation period at two-thirds of the period of revolution. If Mercury's rotation had originally been much more rapid and had been reduced by solar tidal friction, it could finally have "locked-on" when it reached a period near 59 days, exactly in resonance with the revolution period.

We should not be too critical of the visual observers for accepting the 88-day rotation period of Mercury. Such observations are exceedingly difficult. Also, at the times when Mercury is most favorably located for observation, the same face appears several years in a row because of the ⅔ resonance. C. Chapman has studied the old drawings of Mercury made between 1882 and 1963 and combined 130 of them, assuming the precise ⅔ resonance period of rotation and the equator in the orbital plane. His composite map appears in Fig. 142.

The consequence of the ⅔ resonance is to give Mercury an extraordinary long day of 176 Earth days. While Mercury rotates completely with respect to the stars in about 59 days it moves ⅔ of the distance around its orbit. This is $1 - ⅔$ or ⅓ of its day, making the total day 3 times its sidereal rotation period and twice its period of revolution. In Fig. 143 we look down from the pole of the orbit with the body of mercury fixed at the center of the diagram. Choosing the arbitrary zero of longitude at perihelion noon, we see how the Sun appears to move around Mercury in the outer curve. At perihelion Mercury turns slightly slower than it swings around

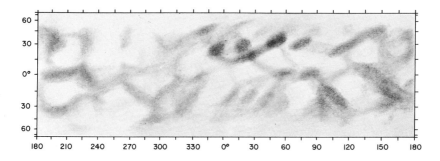

Fig. 142. Map of Mercury. (Courtesy of C. Chapman.)

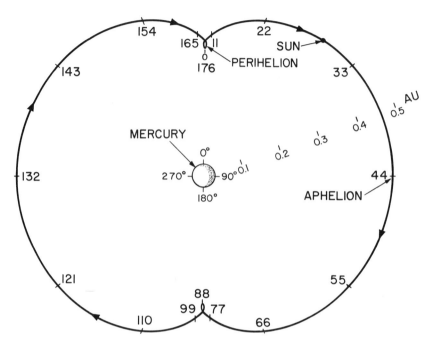

Fig. 143. Mercury's day lasts for 176 Earth days. Projection of solar distance on Mercury's equator with arbitrary longitude marks.

the Sun, so that the Sun moves a little eastward, as seen from Mercury's equator, before it continues its westward daily motion. At longitudes of exactly 90° and 270° on the equator in Fig. 143, the Sun peeks above the horizon for an Earth day or two then sets and rises again to remain ablaze again for nearly a Mercurian year; it then proceeds to set twice. The Sun appears about 50 percent larger in diameter at perihelion than aphelion and delivers more than twice the heat to Mercury's surface. The pole of rotation appears to be within 28° of the perpendicular to the orbital plane according to R. B. Dyce, G. H. Pettengill, and I. I. Shapiro, who determined the radar rotation period.

Evidence for an atmosphere is mostly negative, although the observers have sometimes suspected that faint, whitish clouds obscure the darker markings. The velocity of escape is only 2.6 miles per second and Mercury's surface temperature can rise far above that of the Moon. Hence only extremely heavy gases might be expected to remain on the planet's surface. Also solar storms would tend, even more than for the Moon, to knock away the atoms of a residual atmosphere. The horns of the crescent phase for Mercury do not extend beyond their geometric limits, indicating no appreciable twilight effect of scattering or refraction in the atmosphere. A. Dollfus, however, finds a slight excess in the polarization of light at the cusps. If this is caused by an atmosphere, the total quantity amounts to about $\frac{1}{300}$ of the Earth's atmosphere.

Edison Pettit, of the Mount Wilson and Palomar Observatories, found from infrared measures of Mercury that the temperature of the subsolar point rises at perihelion to the remarkably high value of 780° F and at aphelion to 545° F. At 780° F both tin and lead are molten and even zinc is near its melting point, 786° F. Hence Mercury rather than Pluto might well have been named after the god of the underworld! The exceedingly high temperatures on Mercury's daylight side are to be contrasted with expected low temperatures on the night side. One wonders whether the night side may have trapped and frozen out gases such as nitrogen, carbon dioxide, oxygen, and others. Closer inspection by space probes and radar observations should prove most revealing, although modern infrared and radio measures do not confirm such great temperature extremes as the early measures. The range may be more like $-200°$ F to $+500°$ F, still large but yet uncertain.

The great similarity of Mercury to the Moon has been indicated by its size, sparsity of atmosphere, and superficial appearance. The two bodies reflect light in practically identical fashion with respect both to color and to intensity at various angles of reflection. Light rays falling perpendicularly on the surface are reflected back along their paths with fair efficiency but are reflected at great angles very poorly. Even the polarization, or plane of vibration, of the reflected light is the same for Mercury as for the Moon, and the radar albedo and roughness measures are almost identical with those of the Moon. Hence we may conclude that the surface of Mercury resembles that of the Moon in detail as well as in general character. Certainly the exterior of Mercury is irregular and rough and must be covered with craters.

The mean density of Mercury, now fairly well determined by perturbations for mass and by radar for diameter, is about 5.5 times that of water, or equal to the density of the Earth. Since Mercury's small mass limits its compressional increase in density to only 1 or 2 percent, the mean density of its basic materials, if removed from the planet, would be about 5.4 compared to the

Earth's 4.4, according to calculations by H. C. Urey. Hence Mercury must contain a much larger fraction of the heavier elements than the Earth and, probably, a quite sizable iron core. In this respect, then, Mercury differs markedly from the Moon and represents the intrinsically densest sizable body in the solar system. The evolutionary origin of this high density is not well understood yet but undoubtedly is associated with Mercury's proximity to the Sun.

Mercury's orbit is next to Pluto's in terms of high inclination to the plane of the ecliptic, 7.0°, and eccentricity, 0.21. At perihelion its orbital speed reaches 36 miles per second. Because of this rapid motion and high eccentricity, Mercury has provided one of the three astronomical verifications of the Einstein theory of general relativity. The direction of Mercury's perihelion advances some 43 seconds per century more than is accounted for on the basis of planetary perturbations. Einstein's relativity theory, however, predicts the observed rate within its accuracy of measurement. This remarkable verification of the theory has greatly increased its acceptance in the scientific world. Mercury is thus a major contributor to modern science.

Venus

Venus is both the "evening star" and the "morning star," the Hesperus and Phosphorus of antiquity. It is the most brilliant object in the sky, except for the Sun and Moon. Venus is often visible in the daylight and capable of casting shadows at night. Only 144 days elapse from the evening elongation, when Venus is the first object to be found in the evening twilight, until the morning elongation, when it is the last "star" to disappear in the Sun's morning glow, while 440 days are required for Venus to revolve beyond the Sun and return again to its evening elongation (the geometry can be seen in Fig. 4). Its true period of revolution about the Sun is, of course, much shorter, only 224.70 days. At minimum distance it becomes our nearest planetary neighbor, some 26,000,000 miles distant. But then optical observations are difficult because it appears so close to the Sun.

Venus is truly the Earth's "sister planet," nearly of the same size and mass. The magnification of even a small telescope suffices to resolve the brilliant point of light into a silvery disk, somewhat diffuse at the edges because of the unsteadiness of our atmosphere, but showing the crescent phases like the Moon. When the crescent is thin the horns appear to extend more than half around the disk, as though the irradiation of the brilliant surface were producing an optical illusion. In the extreme situation, however, when Venus lies nearly in the line between us and the Sun, a faint circle of light can be seen entirely around the disk. This twilight arc is shown in the photographs of Fig. 144. Neither our atmosphere nor an optical defect could produce this phenomenon. A deep atmosphere on Venus deflects the sunlight around the edges of the disk by refraction and scattering.

But why do we not see clouds in the atmosphere or else surface markings on the globe itself? Under the best observing conditions, when our atmosphere is clear and steady, only the haziest suggestions of markings can be seen by the most expert observers—"large dusky spots," as Barnard called them. These faint patches, too indefinite to be drawn well, appear impermanent. Photographs in the long wavelengths of infrared light have also proved unsuccessful in registering details, notwithstanding the haze-penetrating power of the infrared light. It was not until F. E. Ross (1874–1960) experimented with the other extreme of the color spectrum, ultraviolet light, that details on Venus could be photographed. To everyone's surprise, because ultraviolet light is markedly useless for clouds on the Earth, Ross was successful in registering great hazy cloudlike formations in the atmosphere of Venus.

Fig. 144. Twilight arc around Venus. Note the extension of the crescent completely around the disk in the first photograph and the bright extension at the top in the second. (Photographs by E. C. Slipher, Lowell Observatory.)

Ross's photographs in Fig. 145 and Lick Observatory photographs in Fig. 146 show that these dark hazy patches change their structures from day to day, proving that they do not represent permanent features of the planet's surface, but are clouds, floating in the sky of Venus. In recent years several observers have found evidence for a four-day retrograde rotation of the Venus ultraviolet clouds, but the reality of this interpretation remains controversial. The spectrograph has been called into use, but even with this instrument the question is not settled.

Only the remarkable feat of bouncing radio waves off Venus provides a solution to the problem of the rotation. The Millstone Hill radar of the Massachusetts Institute of Technology (see Fig. 34), the Goldstone facility at the California Institute of Technology, and a radar in the Soviet Union first succeeded in accomplishing this feat. On page 198 we discussed the radar techniques for measuring planetary dimensions and rotations. For Venus the result is remarkable and at first was difficult to accept. The rotation is *retrograde* with a sidereal period of 243.09 days ($\pm$0.18 day according to I. I. Shapiro), the pole lying within 3° of the orbital pole. A period of 243.16 days would make three rotation periods of Venus resonate with two revolutions of the Earth about the Sun, as though Earth tidal forces might have "locked-on" to the body of Venus. Unlikely as the idea may seem because of the small forces involved, we cannot regard the "locking-on" as impossible. In any case, the slow retrograde rotation is firmly established and theories of planetary evolution must henceforth allow or account for this anomaly in planetary motion. The daily cycle on Venus is 117 Earth days.

Radar shows that the surface of Venus reflects about twice as well as that of the Moon or Mercury, 0.12 of perfect reflection averaged over a wide range of radio frequencies. The surface is significantly smoother, however, than that of the Moon or Mercury; parts of it are very smooth, although the general scattering law (delay curve) is similar to that of the Moon. At least two large reflection features show in the radar echoes. One may be a long mountain chain with height differences measured in miles, first detected at the Goldstone facility of the Jet Propulsion Laboratory.

In Fig. 147 we see the first radar picture of the surface of Venus, obtained by R. Jurgens with the great 1000-foot radio telescope of Cornell University at Arecibo, Puerto Rico. The picture was made as described on page 198 and in Fig. 141. In Fig. 147 *left*, the vertical direction represents the time delay, that is, the distance to the

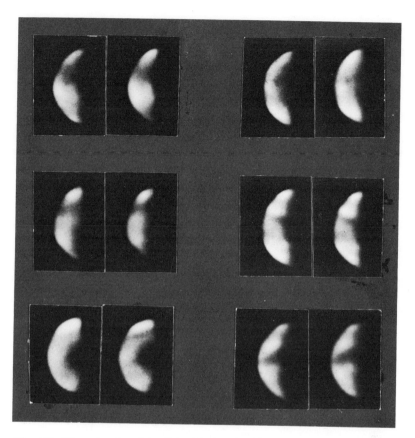

Fig. 145. Venus in ultraviolet light. Each pair of images was made on the same night, the two top pairs on successive nights, and the two middle pairs on successive nights. Note the changes from night to night. The photographic contrast has been increased. (Photographs by F. E. Ross, Mount Wilson and Palomar Observatories.)

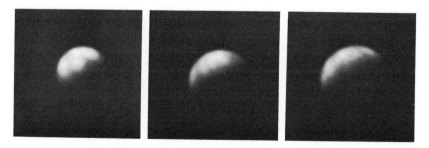

Fig. 146. Venus in ultraviolet light. These photographs were made at the same scale in 1962. (Photographs by the Lick Observatory.)

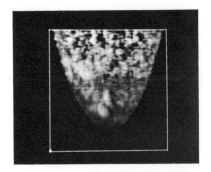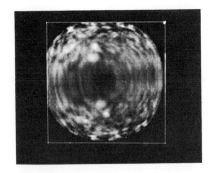

Fig. 147. The first radar pictures of Venus. See text for description. (Courtesy the Arecibo Ionospheric Observatory, Cornell University.)

Earth, including both the northern and the southern half hemispheres but superimposed. The horizontal direction represents radar frequency, that is, rotation or distance from the apparent rotation axis. The scales are not linear in distance so that the nearest area is elongated near the bottom of the picture. The projection is rectified in Fig. 147 *right* to give the appearance of Venus as seen from the Earth, but combining features of both northern and southern hemispheres in the two identical hemispheric projections shown. One of the two conspicuous features in low latitude can be identified with the original Goldstone feature in the northern hemisphere. The reality of a number of minor features requires further confirmation. But no longer can the great clouds of Venus deny us knowledge of her landscape.

Radio techniques have provided an even greater surprise about Venus than its surface features and its retrograde rotation. Venus is very hot! Infrared measures of temperature at the cloud tops have been consistently cool, near −36°F, both on the sunlit and on the dark sides. This value corresponds well to that at the tops of high Earth clouds. But the radio thermal measures shown in Fig. 148 are remarkable. At the very shortest wavelengths, 3 millimeters (0.1 inch), the temperature has begun to rise above the infrared value. At a wavelength of 2 centimeters (0.3 inches) and at longer wavelengths the Venusian temperature has reached the amazing value of more than 600°F. Although there is still some question as to the interpretation of this high temperature, most investigators favor the concept that the longer radio waves are able to penetrate the entire atmosphere and that the true ground temperature may even exceed 600°F! F. D. Drake at the National Radio Observatory in West

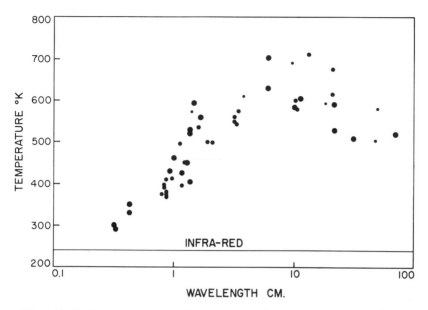

Fig. 148. Radio temperature of Venus as dependent on wavelength (300° K = 81°F, 500°K = 441°F, 600°K = 621°F, 700°K = 801°F). (After D. Barber and H. Gent.)

Virginia finds that the temperature range at a wavelength of 10 centimeters (4 inches) does not change much around the planet; perhaps the total variation is 30° to 70°F. Thus whatever the high temperature signifies, it applies to all parts of the observable planet. After discussing the nature of the atmosphere we shall return to this question of the true surface temperature.

Let us now look at the spectrographic evidence concerning the composition of the atmosphere. No evidence for oxygen has been found. C. E. St. John (1857–1935) and S. B. Nicholson (1891–1963) of the Mount Wilson Observatory demonstrated that the sunlight reflected from Venus passes through less than the equivalent of a yard of oxygen gas at sea-level pressure on the Earth, or less than a thousandth part of the oxygen in our atmosphere. In addition, the amount of water vapor above the clouds on Venus is less than that in our atmosphere by at least a factor of ten. John Strong of Johns Hopkins University studied Venus from high-altitude stratospheric balloons and found some evidence for the presence of water vapor. The amount above the clouds appears not to exceed 0.1 millimeter of liquid water.

W. S. Adams (1876–1956) and T. Dunham at the Mount Wilson Observatory, however, obtained more positive results from their spectrographic analysis. They discovered new infrared absorptions that were unknown from laboratory studies. Calculations indicated that the unknown absorptions might arise from ordinary carbon dioxide, if the light passes through a sufficiently long column of this gas. To check the theory Dunham filled a 60-foot pipe with carbon dioxide compressed to ten times the pressure of our atmosphere. When artificial light was sent down the tube and reflected back to the same spectrograph used for the Venus spectra, identical absorptions were obtained. A spectrogram of Venus is compared with the solar spectrum in Fig. 149. The amount of carbon dioxide vapor above the clouds is enormous compared with that in the Earth's atmosphere; the amount varies but may equal a mile equivalent thickness at sea-level pressure compared to 7 feet for the Earth's atmosphere. Observers at the French observatory of Saint-Michel have found evidence for traces of hydrochloric acid (HCl) and probably hydrogen flouride (HF) in the spectrum of Venus.

H. Spinrad has studied both strong and weak bands of the carbon dioxide absorption spectrum and finds that the weak bands, corresponding to lower levels in the atmosphere, indicate higher temperatures than the strong bands which arise from very high levels. Hence it appears that the observable part of the atmosphere of Venus may be like the Earth's atmosphere below the stratosphere where the temperature decreases with increasing height. Furthermore, the strength of the weak bands proves that sunlight is highly scattered and that the clouds are really just thin haze of great depth.

There is no hope remaining that the surface temperature of Venus may be tolerably low. If the planet had an immensely dense ionosphere, that is, with 1,000 to 10,000 times as many electrons per unit volume as the densest part of the Earth's ionosphere, then

Fig. 149. Spectrograms (a) of the Sun, (b) of Venus; (c) Venus widened. Note the infrared absorptions of carbon dioxide (arrows), strong in the spectra of Venus, but absent in the spectrum of the Sun. (Photographs by T. Dunham; courtesy of Yerkes Observatory.)

the radar reflections and the radio temperature measures might both apply to regions in this ionosphere and not to the solid surface of Venus.

The success of the U.S. spaceship Mariner II (Fig. 150) in approaching Venus to a distance of 21,600 miles on December 14, 1962, and in returning scientific measurements of the planet added strikingly to our knowledge about Venus. The Mariner II, under the responsibility of the Jet Propulsion Laboratory of the California Institute of Technology and the National Aeronautics and Space Administration, radioed back scientific information from the then phenomenal distance of 37,000,000 miles. The magnetic field of Venus was found to be less than a thirtieth that of the Earth. Furthermore Venus showed no measurable Van Allen belt of energetic particles, confirming the lack of a magnetic field.

The radio measurements made by the equipment of A. E. Lilley

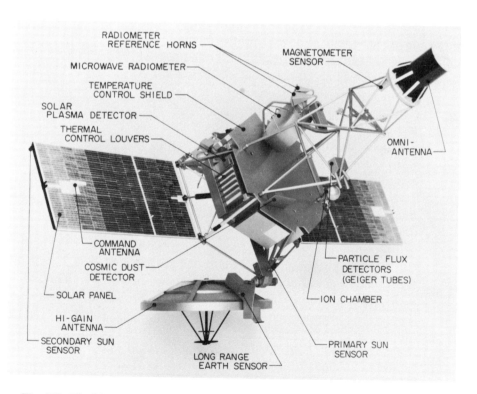

Fig. 150. The Venus spacecraft, Mariner II, that approached Venus on December 14, 1962. (Photograph by the National Aeronautics and Space Administration.)

of the Harvard College Observatory, and A. Barrett of the Massachusetts Institute of Technology, on wavelengths of 13.5 and 19 centimeters, indicate that the temperature as measured by radio is considerably higher near the center of the planetary disk than near the edge. It should be noted here that radio beams at the distance of the Earth completely encompass the disk of Venus and hence give only a mean temperature. At a distance of 21,600 miles, however, the radar antennas were capable of studying areas approximately one-eighth the diameter of the planet to measure the variation of temperature across the disk. Were there a thick ionosphere on Venus, the apparent radio temperature near the edges of the disk should have been equal to or greater than the average. Thus Venus is very hot at the surface. Mariner II also scanned the high atmosphere of Venus in the infrared and found no irregularities across the terminator related to a night-and-day effect. Nor did such an effect occur in the radio measures of the surface. Furthermore, the infrared scanner on Marine II noted no breaks in the clouds, that is, no regions of higher temperature where the scanner might have looked down into the lower and hotter regions. Nevertheless, a cold spot, some 20 Fahrenheit degrees colder than the rest of the atmosphere, did appear near the south end of the terminator. Possibly this cold spot represents a high-level storm or cloud or, conceivably, the effects of some hidden surface feature. Infrared scanning from Earth regularly shows such colder features.

We now have a *direct* temperature and pressure measures at the surface of Venus! In October 1967 the U.S.S.R. Venus 4 spacecraft actually entered Venus' nightside atmosphere at a speed of nearly 7 mi/sec and was slowed down by atmospheric resistance with much heating. It then ejected an 836-pound instrumented canister that parachuted slowly to the surface, radioing back measures of atmospheric temperature, pressure, and composition. The radio cut out abruptly at a maximum temperature of 536°F, which is probably the surface temperature where the canister struck the surface (on a mountain or in a valley?). The slow and irregular descent, 94 minutes for a drop of 16 miles (measured by radar at ejection of the canister), suggests strong turbulent winds, so that the canister was quite likely damaged by wind upon landing, not by the rate of fall. The surface pressure was 12 to 22 atmospheres!

The Venus 4 spacecraft gave an atmospheric composition of carbon dioxide 90 to 95 percent, water 0.4 percent, nitrogen less than 7 percent, and water plus oxygen less than 1.6 percent.

Some 34 hours after Venus 4 landed on Venus in 1967, the U.S. NASA Mariner V spaceprobe circled the lighted hemisphere of Venus at an altitude of some 2500 miles. It detected a carbon dioxide concentration of 72 to 87 percent, in good agreement with the Venus 4 result, but found no evidence of oxygen. Some hydrogen radiation was present at high altitudes and some ionosphere on the sunlit hemisphere, in contrast to none found on the dark hemisphere by Venus 4. The Mariner V placed the magnetic field of Venus at about $\frac{1}{300}$ that of the Earth, again suggesting that the slow rotation of Venus does not make possible internal motions to provide strong magnetic fields as in the Earth.

A high temperature at the bottom of an atmosphere containing a great deal of carbon dioxide is not a very surprising result. The gas is very transparent to all visual light, and, unlike oxygen, to ultraviolet light. Carbon dioxide, however, absorbs heat (far-infrared) radiation extremely well. The result is that the greenhouse effect should be powerful in heating the surface of Venus. Much of the Sun's energy can enter as visual light, while the radiation from the heated surface is trapped by the carbon dioxide. This identical process helps to regulate the surface temperature of the Earth. C. Sagan has shown, however, that a little bit of water vapor can help "seal up the chinks" in the infrared spectrum of carbon dioxide through which heat might escape from the surface of Venus.

On Earth the water acts as a catalyst, enabling carbon dioxide to combine with the silicate rocks, thus "fixing" the carbon dioxide in carbonate rocks. A layer a tenth of a mile deep has been formed around the Earth in geologic times and accounts for about the amount of carbon dioxide now present in the atmosphere of Venus. Thus the surprise in the atmosphere of Venus is not the presence of carbon dioxide but the absence of primitive water.

Again, oxygen in our atmosphere comes from the water dissociated by solar ultraviolet light, the hydrogen being lost to space because of its low molecular weight. Thus the lack of water on Venus prevents the production of much oxygen, in agreement with the observations.

There is evidence, both positive and negative, that the upper clouds on Venus are made of ice crystals. The polarization of light scattered by the atmosphere of Venus can most easily be explained in this fashion and the temperature at the top of the cloud layer seems to be close to $-38°F$, the value at which ice forms under such low pressure. The vapor pressure of ice at $-38°F$ is quite low so that water vapor is mostly removed above Venus' clouds, while the detection of this residual water vapor by observations made through the Earth's atmosphere becomes quite difficult. Most observers, however, fail to find the expected infrared absorption bands if the clouds are presumed to be of ice. The composition of the clouds thus remains uncertain.

A few years after the discovery of carbon dioxide on Venus, Rupert Wildt put forward an alternative explanation for the deficiency of water vapor and the nature of the clouds on Venus, particularly applicable to the now suspected lower layer of clouds. He suggested that hydrocarbons, possibly formaldehyde, might be likely. He noted that, in a mixture of carbon dioxide and water vapor, formaldehyde is formed by the action of ultraviolet light. This manufacture of hydrocarbons can proceed so long as the water supply continues. A certain amount of water is also required for the formation of clouds. Pure formaldehyde gas is colorless and unclouded. The slightest trace of water vapor added to formaldehyde gas, however, instantly produces a thick white cloud. No formaldehyde is observed in the atmosphere of Venus. Other carbohydrates may be possible, however. Kuiper suggests that the clouds on Venus contain carbon suboxide (C_3O_2). No observations, however, support any of these suggestions.

Since the surface of Venus is hot and dry, dust storms should be prevalent and violent. E. J. Opik has suggested that the lower atmosphere of Venus is incredibly dusty and that the very high temperature is maintained more through the action of the dust than because of the greenhouse effect.

Hitherto in our travels through the solar system we have encountered planetary exteriors that are completely undesirable for home sites. The distant planets are too cold; Mercury is both too hot and too cold. The giant planets are covered with noxious or poisonous gases and may possibly lack solid surfaces; Pluto, Mercury, and all but one satellite have no atmospheres whatsoever. Here in Venus we have an Earth-like planet, but it is too hot and has no oxygen. Is Mars the only remaining hope for the possibility of our kind of life on other planets?

13

Mars

Mars was named for the god of war because of the planet's sanguine color, obvious to the naked eye and more conspicuous with a telescope. The name, unfortunately, was much too appropriate during a number of years near the turn of this century. An astronomical battle was raging at that time and Mars was the battlefield. On one side was Percival Lowell, who carried on the banner first raised by Schiaparelli. On the other side stood a considerable fraction of the astronomical world. The *casus belli* was the observation of "canals" on Mars by both Schiaparelli and Lowell, and Lowell's interpretation of these narrow markings as artificial waterways. Schiaparelli used the Italian word *canali*, which means primarily *channels* or *grooves*, and did not believe that the canals were artificial. Lowell based his interpretation on his own extensive observations of Mars. Some of his composite drawings are shown in Fig. 151.

In the scientific world disagreement among authorities contributes to real and substantial progress. Usually the contenders are each

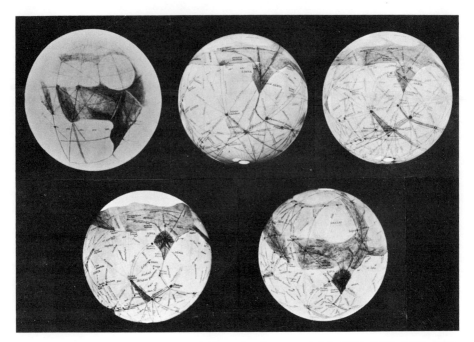

Fig. 151. Lowell's maps of Mars, for the years 1894, 1901, 1903, 1905, 1907, drawn on globes and photographed. (Courtesy of the Lowell Observatory.)

partially right and each partially wrong, but the heat of discussion furthers observation, which is the foundation of science. The Martian battle is over and the smoke has cleared. We can hardly say that either side won a decided victory, but these assiduous observers, seeking the truth, have increased our knowledge of Mars and helped to place its study on a firm basis.

Today we have amazingly powerful methods of studying our planetary neighbors. Great optical telescopes can focus spectrographs and other optical equipment to study the amount, nature, and variations in the light and color of Mars. Great radio telescopes can study the radio emanations, while radars can bounce radio waves from Mars to tell us about its surface, atmosphere, and Van Allen belts, if such exist. Balloons and satellites can carry telescopes above our tremulous, opaque atmosphere for information in the far ultraviolet and infrared unattainable from the ground. Finally, space probes can circle or even land, to radio back firsthand information. But even so, our knowledge is still fragmentary and our true understanding of Mars has increased little since the early triumphs with telescopes collecting light for trained eyes. Fundamentally we must rely upon exploration by interplanetary vehicles to tell us the whole story. Human beings should eventually set foot on Mars as a culmination to man's defiance of nature's greatest barriers, space and gravitation. Before this happy day we must make what we can of the observations we have at hand.

When Mars is most favorably situated for observation, a magnification of some 70 times enlarges the disk to the apparent diameter of the Moon. Small telescopes can be used satisfactorily at such a magnifying power, while larger ones are efficient at much greater powers. Since considerable detail on the Moon is visible to the naked eye, the reader may wonder why Mars should be difficult to observe. The difficulty is again the "seeing," discussed in Chapter 8. Under the best observing conditions on the ground the eye can occasionally resolve contrasting points 20 to 30 miles apart at the minimum distance of Mars. Very large telescopes do only a little better than those of aperture 20 to 30 inches (Fig. 152). Photographs

Fig. 152. Photographs of Mars with the 200-inch reflector: (*left*) in blue light; (*right*) in red light. (Photographs by the Mount Wilson and Palomar Observatories.)

generally resolve no better than about 200 miles although there is some hope that television image tubes, because of their great sensitivity and short exposure times, can catch the flashes of good seeing on a few of innumerable pictures to record the same detail that the eye can see. It is even doubtful that great mountain ranges, such as exist on the Moon or the Earth, could be detected by their shadows on the Martian surface. Such mountains cast no shadows at the nearest position of the planet, opposition to the Sun, and Mars is much farther away at times when such shadows become appreciable. Thus we are not completely certain that Mars is really smoother than the Moon and the Earth, although this statement occurs frequently in the literature.

Even as first observed with a small telescope under very ordinary conditions of "seeing," Mars immediately gains an individuality. Lowell wrote: "Almost as soon as magnification gives Mars a disk, that disk shows markings, white spots crowning a globe spread with blue-green patches on an orange ground." This verbal picture of Mars is somewhat more striking than the sensory registration of a novice who first observes Mars with a small telescope under average conditions. But by persistent observation, night after night, his eye will become more and more expert until he is able to distinguish surface details that were at first completely invisible. This remarkable improvement of visual acuity with experience has sometimes been underestimated, even by skilled observers who have not concentrated on planetary observations.

Three prerequisites are thus essential to a satisfactory study of the surface features on Mars: ideal atmospheric conditions, a "perfect eye," and a first-quality telescope, not necessarily a large one. With these prerequisites the expert observer of Mars must then continue his observations every clear night at every opposition of the planet for several years. Only under these circumstances can he hope to see "all there is to see" on the Martian surface. Naturally there have been few observers who have possessed both the perseverance and the opportunity for exhaustive studies of Mars. Schiaparelli and Lowell provided most of the early visual observations.

Mars can be well observed at intervals of about 2 years and 50 days when it comes into opposition with the Sun. Its synodic period with respect to the Earth is 780 days, about 50 days longer than 2 years (see Appendices 2 and 3 for the geometry and numerical data). Its distance from the Earth at opposition varies by a factor of nearly two (from 34,600,000 to 62,900,000 miles) because of the high eccentricity of the Martian orbit. The most favorable oppositions for observation are, of course, those at which Mars is the nearest, that is, when opposition occurs at perihelion. Since oppositions occur successively later by 50 days in alternate years, a favorable opposition will be repeated in seven or eight periods, at intervals of 15 or 17 years. The perihelion of Mars' orbit is so oriented that Mars is always best located for observation near August (1939, 1956, 1971, 1988). The positions of Mars at various oppositions are shown in Fig. 153.

The equator of Mars, like the Earth's, is tipped some 24° to the plane of its orbit, the direction of the axis remaining fixed in space. Consequently, at the times of Mars' closest approach we always see the planet in the same relative position. The south polar cap, by chance, is the one best observed, the north polar cap being turned toward us at the less favorable oppositions.

Mars rotates on its axis in 24 hours 37.4 minutes, making its day, from noon to noon, nearly 40 minutes longer than our day. The rotation may be noticed after less than an hour's observing. On the succeeding night the planet presents the same side because it

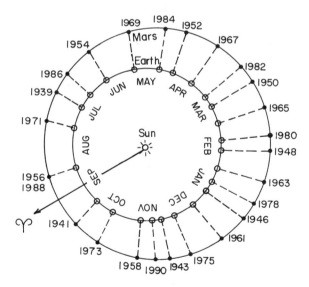

Fig. 153. Oppositions of Mars from the Earth, 1939 to 1990. The relative distances are shown by the lines joining the orbits. The seasonal dates on the Earth are indicated. Mars is north of the equator for oppositions from September to March.

has completed almost one turn in the meantime. In a little over a month the Earth gains a whole rotation, sufficient to complete a cycle of observations entirely around the planet. Similarly, in 24 hours, observers distributed around the Earth can observe the total circumference of Mars under optimum conditions of phase.

Photographically or visually the polar caps are usually the most conspicuous markings on the planet. The seasonal changes, first noted by Sir William Herschel, are regular, and even predictable with considerable accuracy. As the autumn season gives way to winter on one hemisphere of Mars, the corresponding polar cap grows irregularly until it may extend nearly halfway to the equator, to latitude 57° in the northern hemisphere and 45° in the southern hemisphere, the latter being the colder in the winter but the warmer in the summer. With the coming of spring (in the Martian March), the cap begins to recede; by the end of the Martian July it has disappeared at the south pole; the northern polar cap never quite disappears. In Fig. 154, the photographs present the same side of the planet to show the recession of the south polar cap. The seasonal dates for Mars are taken to correspond to seasons on the Earth; it must be remembered that the Martian year is 687 days, nearly equal to 2 Earth years. From a careful inspection of the series of photographs we can see the general darkening around the white area as the cap wanes in March and May, the intensification of the dark areas progressively away from the pole in June and July, and their fading by August. This sequence of events repeats similarly every Martian year.

Even small details in the surface markings will reappear at the same season in different years. A detached area of the south polar cap is visible in the two photographs of Fig. 155. The first photograph was taken in 1909 and the second in 1924, but both on the Martian date June 3. The persistent area bears the name Mountains of Mitchel. There is a question whether it represents a high plateau, a southerly slope, or a depression. The visual observations made by A. Dollfus show considerably more detail about the north polar cap in Fig. 156, where the equivalent Martian dates are June 2 and July 12.

The repetitive character of the changes in the polar caps suggests immediately that these white areas are snow, which melts as the temperature rises. An alternative material is carbon dioxide or "dry ice." We shall return to this question later.

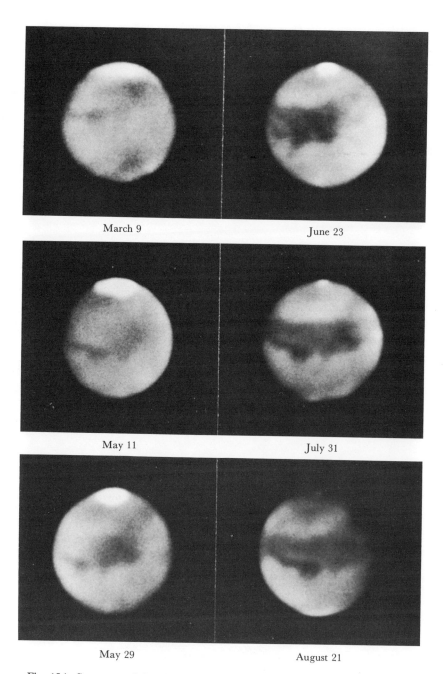

Fig. 154. Seasons on Mars. The dates given are Martian seasonal dates, taken to correspond to those on the Earth. (Photographs by E. C. Slipher, Lowell Observatory.)

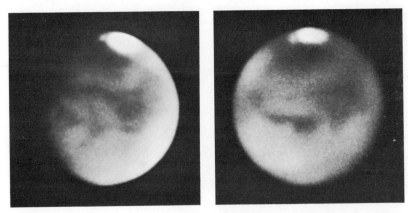

Fig. 155. Mountains of Mitchel. Photographs of Mars in 1909 and 1924. The detached area of the Martian polar cap appears at about the Martian date June 3. (Photographs by E. C. Slipher, Lowell Observatory.)

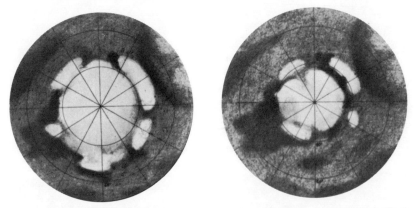

Fig. 156. Development of the north polar cap of Mars, as observed by A. Dollfus with the 24-inch refractor at the Pic du Midi. Equivalent Martian dates: (*left*) June 2; (*right*) July 12. (Courtesy of A. Dollfus.)

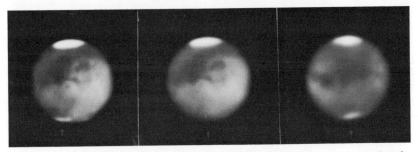

Fig. 157. Clouds at the north pole of Mars, 1939: (*left*) cloud near the north pole, bottom; (*middle*) the cloud has vanished on the next night; (*right*) another cloud 6 days later. (Photographs by E. C. Slipher, Lowell Observatory.)

As a polar cap begins to form in the Martian autumn, variable bluish-white clouds can be observed. The first two photographs in Fig. 157 were taken on successive nights in 1939 by E. C. Slipher, who went to South Africa to observe the planet at its close approach because Mars could be observed more nearly overhead from the southern hemisphere. A white cloud near the north pole in (*a*), bottom, has disappeared by the next night (*b*). Another cloud is present six nights later (*c*). These clouds persist during the growth of the polar caps and special observing techniques are required to distinguish them from the frost or snow on the ground.

Infrared light can penetrate haze and dust in the Earth's atmosphere where blue or violet light will be stopped. Figure 158 presents photographs taken by W. H. Wright (1871–1959) at the Lick Observatory. At the left are Mars and the valley of San José as photographed by violet light, while on the right infrared light

Fig. 158. Mars and the valley of San José, as photographed from Lick Observatory. Violet light was used in the upper photographs and infrared light in the lower ones. (Photographs by W. H. Wright, Lick Observatory.)

has been used. These photographs and those of Figs. 152 and 159 tell their own story. There can be little question that Mars is covered with a hazy atmosphere.

The variable white markings, such as the ones in Fig. 157, are bright in violet light and invisible in the infrared. They are probably thin clouds of carbon dioxide or ice crystals that reflect the violet light but transmit the infrared. Since the semipermanent polar caps are photographed in both colors but are brighter in violet light, they must be actual surface deposits, accompanied by hazy or mistlike clouds. The top three photographs of Fig. 159 illustrate the phenomenon. Note the brightness of the polar cap in the violet as compared with the infrared. Measures of the polarization of the scattered light confirm these conclusions. Dollfus finds that the polar caps are always covered with a veil of thin clouds, probably consisting of fine crystals, perhaps like our cirrus clouds.

Clouds like those seen near the polar caps are frequently observed on other parts of Mars' disk. These bluish clouds persist for only a few hours. The formation of such a cloud during a Martian afternoon is depicted by the series of photographs in Fig. 160. The images are duplicated with identification arrows. As Mars rotates through about 55°, the cloud, which is not present in the first

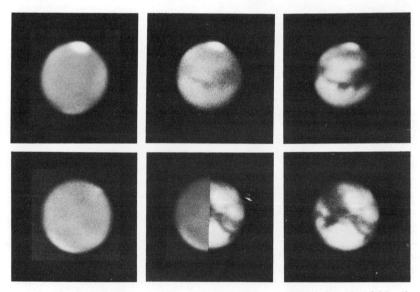

Fig. 159. Mars in various colors: (*upper left*) violet; (*middle*) infrared; (*right*) yellow; (*lower left*) ultraviolet; (*middle*) half ultraviolet, half infrared; (*right*) infrared. (Photographs by W. H. Wright, Lick Observatory.)

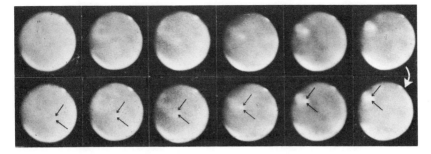

Fig. 160. A Martian afternoon. The arrows indicate a cloud that developed on the successive photographs. Mars turned toward the left during the 4-hour interval. The two rows of images are identical except for the arrows. (Photographs by the Lick and Mount Wilson and Palomar Observatories.)

photograph, is seen to develop and traverse the disk. It is brightest near sunset.

Some of these extremely thin, bluish clouds, usually invisible in orange and red light, are almost always present at the equator near sunrise and sunset, disappearing in the middle of the day. Dollfus regards them as morning and evening tropical fogs, sometimes connected with the denser formations of white clouds. Also he finds from the polarization that the bluish clouds are situated higher in the atmosphere than the white clouds and seem to be comparable with the *mother-of-pearl* clouds that are occasionally photographed in blue light in the terrestrial polar regions near an altitude of 20 miles.

The fact that the diameter of Mars appears greater in the violet than in infrared light (see Fig. 159) has been explained as an effect of haze in the atmosphere. However, the difference in depth—some 40 miles—is much too large for this explanation. With greatly improved techniques Dollfus finds no such effect.

The atmosphere of Mars is certainly hazy. The remarkable characteristic of the haze is the fact that it occasionally disappears, permitting the surface features to be photographed in blue light as well as in orange, yellow, or red light. G. de Vaucouleurs, however, has shown that the "blue clearing" on Mars can also occur at other times and that possibly its more frequent observation near opposition is associated with increased observing at that time.

A great deal of discussion concerning the blue clearing has not entirely explained the nature of the haze. Öpik points out that its designation as "blue haze" is inappropriate. The haze does not, indeed, reflect well in blue light; the albedo of the planet is nearly 0.3 in the deep red, about 0.15 to 0.05 in yellow light, and only 0.04 in the deep blue. E. C. Slipher finds that frequently there are extremely dark areas of large extent in the blue photographs. Öpik supports the concept that the haze is truly "red haze," at least from the viewpoint of any Martians, because it would transmit red light better than blue light. He attributes much of the reddish color of Mars to the existence of this haze, which reddens the light from the Sun as it passes through the Martian atmosphere and again as it passes back to the Earth. Thus the Martian deserts may be only slightly less drab in color than the surface of the Moon.

The possibility that the "red haze" may be of organic origin is highly tempting; it should neither be believed nor forgotten at the present time. Incidentally, Öpik suggests that the haze may be more opaque to near-ultraviolet light than our atmosphere, so that one might sunburn less on Mars than on the Earth.

Mars shows another anomaly that may possibly be associated with its atmospheric haze. The observed oblateness of the planet, that is, the difference between the equatorial and the polar diameters divided by the equatorial diameter, is $\frac{1}{80}$, while its calculated value, based on the motions of the satellites, is only $\frac{1}{190}$. The observed value requires a polar flattening of 26 miles in radius and the calculated one only 11 miles, a difference of 15 miles. To account for this difference by assuming lighter-density high plateau regions near the poles would require a plateau region 15 miles high, which is completely out of the question. The effect may be partially explained by a thinner haze layer at the polar regions and possibly by other complicating effects such as contrast in brightness near the limb and systematic errors entering into the techniques of measurement at different wavelengths.

Besides the white and bluish clouds there are occasional yellow clouds, invisible in violet and ultraviolet light. The yellow clouds may persist for several days. They appear to occur low in the Martian atmosphere where the violet light does not penetrate. Most observers interpret the yellow clouds as dust. On rare occasions a large fraction of Mars' surface is covered with such dust. In fact the close opposition of 1956 was marred by a persistent dust storm that hid most of the surface features. It has been suggested that in rare cases such dust clouds may have been initiated by large meteorite falls.

If the "blue" clouds are mist and the polar caps are snow, which is carried by the atmosphere from pole to pole during alternate seasons, we should expect water vapor to be observable in the Martian atmosphere. Very thorough studies failed to indicate the slightest trace of water until finally, in April 1963, H. Spinrad, G. Münch, and L. D. Kaplan, with the Mount Wilson 100-inch reflector, obtained an infrared spectrogram showing water-vapor lines in the light from Mars. Shifted by a radial velocity of 10 miles per second from the lines produced in the Earth's atmosphere, the lines were strongest over the Martian poles, corresponding to about 0.0004 inch of liquid water above the surface. Other observers find the water vapor variable from time to time if present at all. Probably 0.001 inch equivalent of liquid water is near the maximum value possible.

The test for oxygen still gives negative results, less than 0.1 percent of the Earth's. There is enormously more oxygen above Mount Everest than above the surface of Mars.

A number of other gases, including ozone (O_3), methane (CH_4), ammonia (NH_3), nitrous oxide (NO_2), and carbon monoxide (CO), have been proved to be absent or of extremely low abundance in the atmosphere of Mars. In 1952, however, Kuiper positively identified carbon dioxide in the spectrum. The amount is some 14 times that in the Earth's atmosphere and corresponds roughly to a 200-foot layer at the Earth's surface.

The spectacular U.S. spacecraft Mariner IV of the National Aeronautics and Space Administration instrumented by the Jet Propulsion Laboratory sailed past Mars at a distance of 6118 miles on July 14–15, 1965. As the spacecraft passed behind the disk of Mars, the radio signals were bent by the atmosphere of Mars. Measurements of the rate of fadeout could be interpreted in terms of the total atmosphere present. The result turned out to be disappointingly small, only 0.8 percent of the atmospheric density near the Earth's surface. Thus the measured CO_2 content must constitute a major fraction of the atmosphere, while perhaps nitrogen and argon supply the remainder with a trace of water vapor. Even though the gravity of Mars is only 0.38 that on Earth so that the atmosphere is less compressed, the atmospheric density never reaches that on Earth at any corresponding altitudes. G. and W. C. Fjeldbo and V. R. Eshleman find that the temperature of the atmosphere near the surface is extremely low, about $-135°F$, and falls with increasing altitude. Thus carbon dioxide haze or snow crystals might condense out at moderate heights. The carbon dioxide should be dissociated by solar ultraviolet light at altitudes of the order of 50 miles, so that the higher atmosphere is largely molecular oxygen.

The sparsity of atmosphere on Mars is disappointing from the viewpoint of spacecraft landings because it probably will preclude the use of parachutes. Reverse rocketry may be required, reducing greatly the effective payload of the already expensive vehicles and complicating the remote-control technology of soft landings.

The temperatures on Mars, as measured by heat radiation in the infrared, show extreme ranges, from above 70°F at noon near the subsolar point to a sunrise minimum well below −100°F. Fig. 161

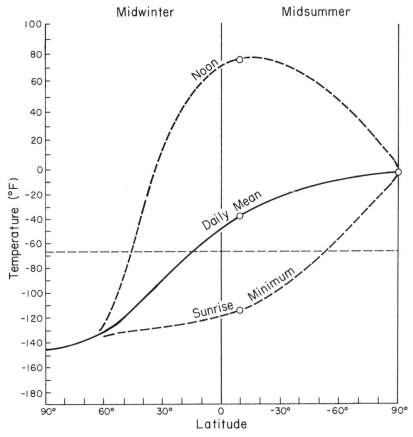

Fig. 161. Mean and extreme surface temperatures on Mars, plotted as a function of latitude for midwinter in the northern hemisphere and midsummer in the southern. The circles represent the best determinations. (After Y. Mintz.)

presents typical values collected by Y. Mintz for the northern hemisphere in midwinter and the southern hemisphere in mid-summer. The minimum measured sunrise temperature of $-110°F$ was made by W. M. Sinton and J. Strong; earlier measures were made by W. W. Coblentz and C. O. Lampland at the Lowell Observatory and by E. Pettit and S. B. Nicholson at the Mount Wilson Observatory. It is possible that even lower temperatures occur late in the Martian night. Thus the total diurnal temperature range may even exceed 200 Fahrenheit degrees.

Calculations based upon the solar radiation and albedo of Mars indicate that its mean temperature should be around $-65°F$. Radio emission (wavelength 0.3 to 21 centimeters) from Mars, measured by many observers, yields a mean temperature near $-90°F$, which may be an indication of the average temperature a short distance below the surface of Mars. At least, the fair agreement between the radio measure and the calculated value suggests this conclusion. Even a fairly balmy temperature of $70°F$ at midday on an optimum location on Mars does not mean that the air should be comfortable there. Mintz calculates that the atmosphere a short distance above the ground might be colder by 80 degrees than the actual ground temperature because the atmosphere is such a good radiator and the greenhouse effect is small. The Mariner IV data gave an even lower temperature. Thus Mars is not so much more attractive for human habitation than is the Moon.

Öpik calculates that light gases such as water vapor and methane should escape from Mars fairly rapidly because of the low velocity of escape, 3.1 miles per second, and since the carbon dioxide would be dissociated by solar ultraviolet light, the oxygen, too, should disappear. He believes that the water vapor and carbon dioxide must be replaced by gases seeping up from the inside of the planet, volcanic in origin but not necessarily involving volcanic eruptions. Meteoritic falls, of course, may add some gases.

Because of the very low temperatures and atmospheric pressure on Mars, the polar caps may well be carbon dioxide snow, mixed with a bit of water frost. Possibly the white clouds and mists are also mostly carbon dioxide.

Dollfus has studied the scattered light from Mars with extreme thoroughness, using all the methods possible in photometry and in polarization studies. He finds only two substances that show the unusual scattering properties exhibited by the Martian surface. These are the minerals limonite and goethite, both of which are iron oxides consisting of Fe_2O_3, the limonite containing two oxide molecules to three of H_2O, and the goethite one oxide molecule per H_2O molecule. He expects to find the surface a light dry dust with a considerable fraction of these oxides present. The indicated presence of fairly frequent dust storms is consistent with this concept. Many investigators have searched for other materials with the properties observed on Mars. They almost universally agree with Dollfus. Underneath the surface, however, we should expect the entire planet to contain permafrost, because of the low temperatures and the indicated presence of some water.

The reality and permanency of the markings on Mars can be confirmed by the reader after a careful comparison of the various photographs and drawings in this chapter. The series of photographs in Fig. 162 show the planet turned successively about 30°. The ease

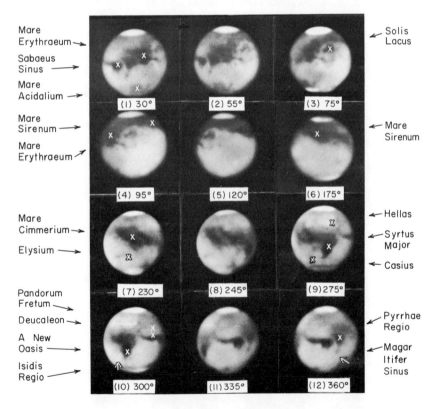

Fig. 162. A panorama of Mars. These photographs of Mars in 1939 show the planet turned through successive angles of about 30°. (Photographs by E. C. Slipher, Lowell Observatory.)

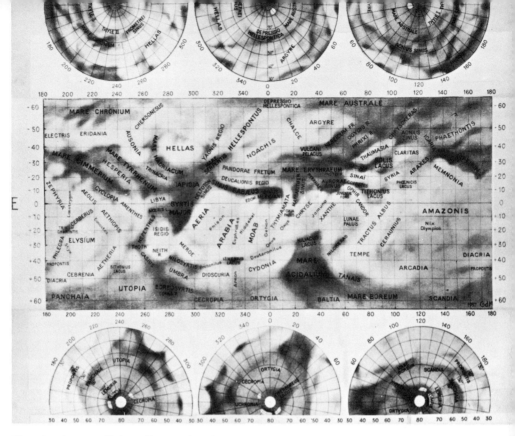

Fig. 163. A map of Mars. (International Astronomical Union.)

with which the markings can be followed from image to image demonstrates their reality and is a tribute to the excellent quality of these photographs by E. C. Slipher. Fig. 163 is a detailed map of Mars. The more conspicuous surface features are identified by their Latin names. *Mare* means a sea, *sinus* a bay or a gulf, *lacus* a lake, *lucus* a grove or wood, *fretum* a strait or channel, and *palus* a swamp or marsh. Actually, there are no extended bodies of water on Mars, the names having been given, as for the Moon, by analogy with terrestrial features. A body of water would reflect the sunlight as a bright point. Notwithstanding careful search, no solar-reflection point has ever been observed on Mars.

The aquatic names refer to darker areas. The brighter regions (exception: the polar caps) are the great deserts. These regions appear ochre or orange in the telescope and generally have unqualified names, such as Elysium, Electris, Hellas, and so on. The deserts give Mars its red color. The darker areas are described as green, blue-green, or gray in color, with many variations of these shades. Various observers agree rather well in assigning colors to the different regions but most modern students of Mars agree that the dark areas are, in fact, chiefly shades of gray, the apparent greenish tints arising largely from a physio-optical illusion. This conclusion does not imply psychological illusions in the careful observations by other observers of Mars. It involves rather the complicated and poorly understood process of color vision.

The seasonal changes in the dark areas of Mars are striking. Even though the color changes are largely illusion, we keep for the record Lowell's observations of color changes in Mare Erythraeum in 1903, shown in Table 2.

Those who see colors on the surface of Mars agree that the dark areas change seasonally from blue-green in the spring and summer to brown or chocolate brown in the midwinter. The seasonal changes are fairly well repeated each year but certain dark areas may be conspicuous for a few oppositions and largely disappear at other oppositions.

The acrimonious controversy concerning the canals on Mars is now settled. Everyone agrees that such markings can be seen and that there are numerous dark spots at their intersections, commonly called oases. These markings, hardly classifiable as a network, change in visibility with the Martian seasons and, to some extent, at the same season from year to year. They become more visible in early summer as the dark areas intensify. No one agrees with Lowell that these are waterways, artificial or not. Most of the features in old drawings can be identified in new photographs taken at the corresponding Martian dates. Sometimes, as E. C. Slipher points out, the photographs show more features than the drawings. To most observers the surface of Mars, when glimpsed at the rare moments of perfect seeing, is much too complex in structure to be represented by a geometric pattern of lines. Many observers do not see the canals as straight lines but generally as curved. It seems

TABLE 2.   *Color changes on Mars as recorded by Lowell in 1903.*

Martian date	Aspect	Martian date	Aspect
Dec. 27	Blue-green	Feb. 17	Faint chocolate
Jan. 16	Blue-green	Feb. 19	Faint blue-green
Jan. 31	Chocolate	Mar. 6	Faint blue-green
Feb. 4	Chocolate	Mar. 8	Faint blue-green
Feb. 13	Faint chocolate	Mar. 23	Pale bluish-green

clear that few if any "canals" are straight and that a variety of structure and width is manifest. The existence of a continuous network is questioned. The difficulties of the problem are best illustrated in two drawings made by W. H. Pickering and reproduced in Fig. 164. At a distance of several feet, where the eye cannot distinguish fine detail, the two drawings look quite similar. Thus the canals appear to be chance markings.

Mariner IV's outstanding feat of actually telemetering back photographs of the Martian surface now gives us a completely new understanding of Mars. Fig. 165 shows the region of the Mariner IV photographs, almost at the ends (near longitude 180°) of the map in Fig. 163.

The first two Mariner pictures (Fig. 166) cover large areas in broad prospective near the limb of the planet as seen from the spacecraft. Bright and dark areas appear as mottling and may include clouds as well as surface markings. Picture 3 (Fig. 167), of a desert region including a "canal" (Fig. 165) on the south edge of the region, shows rill-like structures and probably some craters (Sun zenith angle 14°). In Picture 4 the "canal" crosses the center. Indeed the area *is* dark. Detailed interpretations are difficult to assess.

Mariner Pictures 4 and 5 at high Sun (Fig. 166) resolve details comparable to Picture 6 (Fig. 168), mostly in the desert Zephyria. Small craters are almost certainly present, but the other bright and dark areas are not obviously interpretable. Are they like dark lunar maria and brighter highlands?

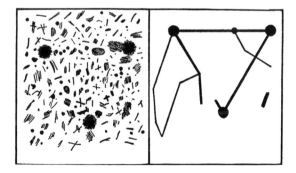

Fig. 164. Stand several feet from this figure and note the similarity of the two diagrams. The canals on Mars may be curves, straight lines, or a succession of smaller markings. (From W. H. Pickering, *Mars;* courtesy of Richard G. Badger.)

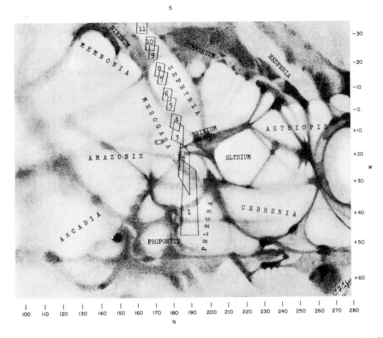

Fig. 165. Photovisual map of the Mariner IV. Scan region on Mars. (C. F. Capen)

Picture 7 at Sun zenith angle 29° (Fig. 168), is the first to show a large number of clearly identifiable craters along with dark and light markings reminiscent of lunar maria and highlands as well as some probable rills. Several of the craters have flat floors and over-lapping appears frequently. Several small mounds or mountains are visible. It is of great interest that Picture 7 lies just within the edge of a region that is dark at some Mars oppositions. Picture 8 (Fig. 168) contains larger craters, up to nearly 40 miles in diameter, and white arcs. Picture 9 at Sun zenith angle 38° (Fig. 168) presents several craters with diameters of the order of 20 miles, one with a central mountain, and a number of smaller craters. The area is mostly in Mare Sirenum. Picture 10 (Fig. 169) covers parts of two 70-mile craters as well as many smaller ones and some sizable arc structures. Does a "ray" structure or a mountain ridge cross the left central area?

The most spectacular picture of the Mariner series is No. 11 (Fig. 170), at Sun zenith angle 47°, showing a great arc of a 100-mile or greater (double?) crater, many smaller craters, including

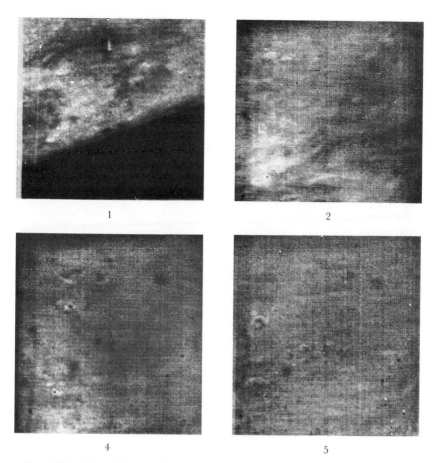

1   2

4   5

Fig. 166. Mars. Mariner IV pictures 1, 2, 4, and 5, south up. No. 1: 410 mi. along limb; 800 mi. A to limb; orange filter. No. 2: 290 mi. EW; 530 mi. NS; green filter. No. 4: 210 mi. EW; 270 mi. NS; orange filter. No. 5: 190 mi. EW; 220 mi. NS; orange filter. (Courtesy U.S. National Aeronautics and Space Administration.)

some perched on larger crater rims, great crooked rills, mountainous areas, and many finely detailed structures. The region is reminiscent of the partially submerged lunar craters on the edges of lunar maria. Most of the large features appear to flatten towards the right in the picture as though the whole region were tilted toward submergence.

Picture 12 (Fig. 169) seems to resolve less detail than No. 11, although the region itself may be duller. Picture 13 (Fig. 169), besides large, bright-rimmed craters, includes a bright arc of very

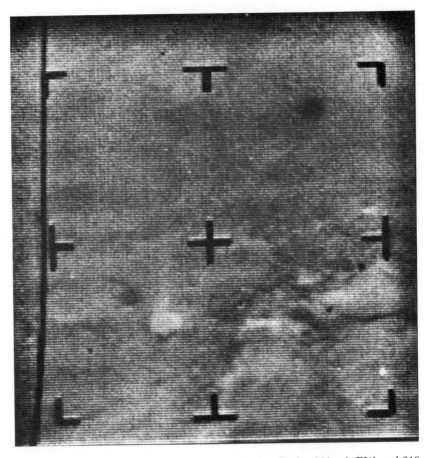

Fig. 167. Mars. Mariner IV picture 3, south up. Scale: 220 mi. EW and 310 mi. NS, green filter. Fiduciary marks are evident, including black circles towards upper right and lower left. (Courtesy U.S. National Aeronautics and Space Administration.)

small curvature, possibly 600 miles in diameter if completed. Picture 14 (Fig. 169), with the Sun zenith angle now at 60°, shows mountains, craters, contrasting dark areas, and much interesting detail. Is there frost on some of the mountain tops at latitude 42° S? Note the large overlap on the lower right with Picture 13. Picture 15 (Fig. 171) shows a few craters and contrasting light and dark areas. The remaining seven pictures typified by No. 16 (Fig. 171) give little or no detail because of technical difficulties.

In comparing the Mariner pictures one notes that those with the green filter generally resolve more detail than contigious ones

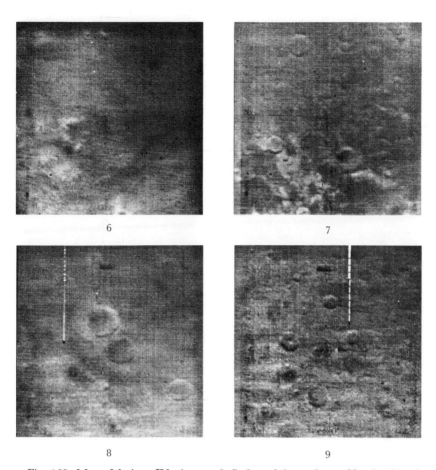

Fig. 168. Mars. Mariner IV pictures 6, 7, 8, and 9, south up. No. 6: 190 mi. EW; 200 mi. NS; green filter. No. 7: 180 mi. EW; 180 mi. NS; green filter. No. 8: 180 mi. EW; 170 mi. NS; orange filter. No. 9: 170 mi. EW; 160 mi. NS; orange filter. (Courtesy U.S. National Aeronautics and Space Administration.)

with the orange filter; the difference is most striking in the overlap region of 13 and 14 (Fig. 169).

The major conclusion to be reached from the remarkable Mariner pictures is that a half-million square miles of Martian surface looks amazingly like the surface of the Moon when viewed at a resolving power of 2–5 miles. To duplicate such views look at half-inch square sections of Fig. 96, well out of focus. The sculpturing by impact craters was to be expected because of Mars' proximity to the asteroid belt. The influx of large meteorite bodies should

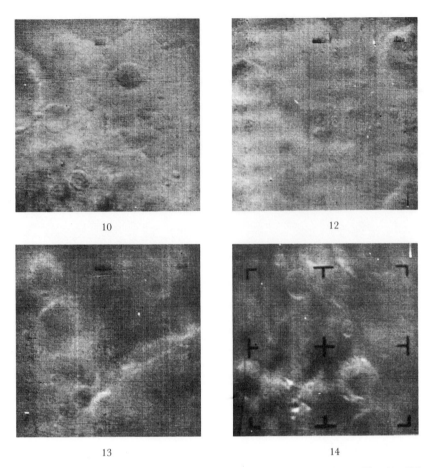

Fig. 169. Mars. Mariner IV pictures 10, 12, 13, and 14, south up. No. 10: 170 mi. EW; 160 mi. NS; green filter. No. 12: 170 mi. EW; 150 mi. NS; orange filter. No. 13: 170 mi. EW; 140 mi. NS; orange filter. No. 14: 170 mi. EW; 140 mi. NS; green filter. (Courtesy U.S. National Aeronautics and Space Administration.)

exceed the rate on the Earth or Moon by at least one order of magnitude if not two. The scars on the Earth are largely healed by mountain building, tilting, and erosion on the Earth over intervals of some 10^7 years. The Moon preserves a record of most craters 3 miles or more in diameter for a much longer time. We do not know yet just when the Moon was formed, but probably not less than 2×10^9 years ago, possibly concurrently with the Earth. The oldest 3-mile craters *could* be more than 4×10^9 years of age.

Fig. 170. Mars. Mariner IV picture No. 11, south up. Scale: 170 mi. EW, 150 mi. NS; green filter. (Courtesy U.S. National Aeronautics and Space Administration.)

On Mars we have no knowledge yet of the healing rate for impact craters. The number of craters observed could accumulate in a minimum of some 10^8 years, probably in a few times this period. Thus obliteration by dust and by whatever plutonic processes occur on Mars requires a longer time than that; 5×10^8 years or longer? The interior probably does not have an active internal circulation, because the Mariner IV measurements show that the magnetic dipole field of Mars, if any exists, does not exceed ⅓₀₀₀ the strength of the Earth's. Thus we have no basis for estimating the plutonic activity on Mars, although the late D. B. McLaughlin proposed that the surface markings may be the consequence of volcanoes

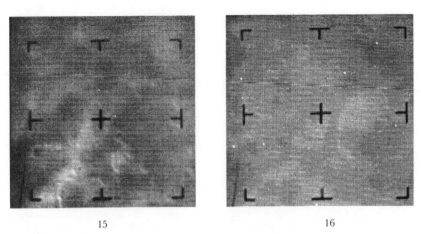

15   16

Fig. 171. Mars. Mariner IV pictures 15 and 16. South up. No. 15: 180 mi. EW;
140 mi. NS; green filter. No. 16: 190 mi. EW; 140 mi. NS; orange filter. (Courtesy
U.S. National Aeronautics and Space Administration.)

whose gases and ashes blow with the prevailing winds and thus
produce the persistent but variable dark regions that we see.

The minor deviations from the normal darkening at corresponding seasons of different years, similar variations in the visibility of
the canals and oases, and the general variability in visibility of
detailed markings are all suggestive of vagaries in the Martian
weather. Almost any part of the surface shows variations of one
kind or another. The generally disappointing 1956 opposition led
E. C. Slipher to a new discovery about Mars. He observed that frequently the regions of greatest cloudiness soon became regions
of extraordinary darkness on red photographs. In other words,
some physical reaction related to the clouds turned the soil very
dark over large areas. We cannot accept the explanation of actual
rain or precipitation because of the thin, dry atmosphere, but
we can admit some type of surface reaction with the clouds, or
some effect of wind-blown dust. On the basis of radar reflection
from Mars, C. Sagan and J. Pollack suggest that the dark areas
are *higher* than the desert regions, the differences and variations in
brightness arising from the wind's ability to carry fine dust to
greater altitudes than coarser dust.

G. H. Pettengill, using the powerful radar system of the Lincoln
Laboratory of the Massachusetts Institute of Technology, finds that
in latitude 21°N on Mars the altitude differences between the

highest mountain tops and the lowest valleys are some eight miles. Thus the gross roughness on Mars compares with that on Earth from mountain tops to the bottom of ocean basins. He finds no strong correlation of altitude with dark areas or desert markings on Mars. Again we await complete radar pictures of Mars to clarify fundamental problems concerning the true nature of the Martian topography.

The temptation to conclude that the dark areas actually contain vegetation is almost irresistible; before coming to such an important conclusion we must look more critically at the evidence at hand. In 1959 W. Sinton discovered in the spectrum of Mars some infrared bands that he tentatively identified with absorptions by natural organic material such as wood, leaves, etc. Unfortunately, these identifications cannot be supported and we are left with *no direct evidence* of organic material on Mars. But the lack of evidence for life in the Mariner pictures is not disturbing. Pictures of similar quality on the Earth would have only a minute probability of proving man's presence here. Over 300,000 pictures of the Earth from NASA's Tiros satellites were made before artificial markings, mile-wide swaths through forests of Canada, could be detected.

Indirect evidence that vegetation conceivably exists on Mars is plentiful: the temperature range is not quite impossible; the polar caps may well contain frozen water; clouds occur that might be made of ice crystals; the wave of quickening from the edge of the melting polar cap suggests the growth of vegetation; the changes in shade (and color?) of the dark areas and canals are favorable; and even the observed irregularities in these changes would be expected. In addition, Dollfus finds characteristic variations with season in the polarization of light scattered from the dark areas, but no such changes from the desert areas. Öpik also argues that only resurgent vegetation could account for the fact that the dust storms do not completely cover the dark areas of the planet to leave the surface essentially uniform.

Furthermore, we have indications that the complex process of life can occur under favorable circumstances without miraculous intervention. For early and penetrating thought on this subject the reader is referred to the Russian biologist A. I. Oparin (*The Origin of Life*, Macmillan, New York, 1938). In 1953 Stanley L. Miller at the University of Chicago, testing a theory proposed by H. C. Urey, conducted what is now a classical experiment. He sealed the simple substances ammonia, hydrogen, methane, and water in a tight vessel, applied some heat, sent electric discharges through the material, and produced a number of the complex amino acid molecules, which are essential to the life processes. In 1963 a significant later step toward the laboratory generation of life was made by C. Sagan, C. Ponnamperuma, and R. Mariner. Starting with complex molecules generated by Miller and his successors, they subjected certain of them to ultraviolet light to produce the compound adenosine triphosphate. This substance, usually designated ATP, is "the 'universal' energy intermediary of contemporary terrestrial organisms, and one of the major products of plant photosynthesis." The ultraviolet light used could not be transmitted through our present atmosphere because of the oxygen and ozone present. But the expected early atmosphere with an excess of hydrogen should have been adequately transparent.

Hence it is essentially proved that complex organic molecules, precursors of living organisms, could have developed by natural causes on the Earth. There is now little reason to doubt that life developed spontaneously on the Earth. Less faith is required to believe in this explanation for the origin of life than in the other more subjective explanations. Similar processes should have occurred on Mars. Various investigators have shown that a few terrestrial microorganisms can *survive* under the rigorous temperature changes, the low atmospheric pressure, the dearth of water vapor, and the absence of oxygen that we know to exist on the surface of Mars. After a reasonable development of primitive "life" molecules on Mars, the Darwinian principle of natural selection would then operate there as well as here to produce more and more complex systems in biological evolution. It seems not unlikely that the early history of Mars may have paralleled in many respects the early history of the Earth, providing a reasonable probability that biological evolution may have proceeded there, but to an unknown degree.

To continue our happy speculation, we might expect the plant life on Mars to resemble desert or high-altitude vegetation on the Earth. Growths would probably be sparse and must be exceedingly resistant to subzero temperatures. Desert plants that can survive by storing water for long periods of time might adapt themselves to the rigorous Martian environment. Mosses and lichens seem the most likely plant life, since they are the last terrestrial plants to disappear from mountain slopes and arctic tundras. Lichens, however, may be too complex to have developed under the difficult conditions on Mars.

Although little oxygen and water now remain on Mars, they may well have been abundant in the distant past.

Urey has developed strong evidence for his theory that in the early stages of formation of the Earth, and presumably of Mars, the atmospheres were reducing rather than oxidizing, that is, they contained a considerable portion of free hydrogen mixed with such basic simple compounds as ammonia, methane, and water. The hydrogen was then lost from the high atmosphere by dissociation processes and finally the atmosphere became oxidizing, that is, free oxygen was present. The red (?) deserts of Mars may tell the story of the lost oxygen, which combined with the iron of the rocks to produce the deserts we now see. The oxygen may have literally *rusted away*. On Earth it appears that volcanic action has supplied the major portion, if not all, of the present oceans and atmospheres. Lesser volcanic activity and lower surface gravity on Mars may account for the present difference in the surface and atmospheric characteristics of the two planets. The low velocity of escape, only 3.1 miles per second, must permit an exceedingly slow but appreciable loss of the atmosphere. The outermost cold layers of the planet absorbed rather than augmented the supply of gases and liquids on the surface. A large quantity of water may be held in chemical combination with the rocks, in the form of hydrates, as well as in the form of permafrost. In such a fashion Mars may have aged, to become the dying world that we observe today.

If we have correctly reconstructed the history of Mars, there is little reason to believe that the life processes may not have followed a course similar to terrestrial evolution. With this assumption, three general possibilities emerge. Intelligent beings may have protected themselves against the excessively slow loss of atmosphere, oxygen, and water, by constructing homes and cities (small or underground, else they would have been observed) with physical conditions scientifically controlled. Strange, though, that they have no radios. As a second possibility, evolution may have developed a being who can withstand the rigors of the Martian climate. The race may have perished, conceivably by a nuclear holocaust. Or, most likely, evolution stopped at some early stage. These possibilities have been sufficiently expanded in the pseudo-scientific literature to make further amplification superfluous. That intelligent beings other than ourselves exist to appreciate the splendors of the Martian landscape is unsupported optimism.

To come back to reality, we must note that the greenish colors of the dark areas are not substantiated by modern observers, that even the redness of the deserts *may* be caused by haze, and that the seasonal variations probably represent some type of nonorganic interaction of or with surface materials. Also the surface structure appears now to have been largely determined by the infall of great meteorites from the asteroid belt, as suggested by Öpik and C. Tombaugh. The oases may be splendid examples of large meteor craters. The polar caps and white clouds may be carbon dioxide snow.

Speculation about life on Mars is a stimulating mental exercise. It may be succeeded by greater speculation as space probes bring us more detailed information about the nature of the planet. No more challenging problem faces us than the origin and nature of life. Do we have living neighbors, even if only primitive plants and animal structures in these near reaches of space? Mars gives us almost the only remaining hope that this may be true unless the unexplored surfaces of the giant planets harbor the unexpected.

14

Origin and Evolution
of the Solar System

No longer can the philosopher in his easy chair expect to solve the basic problems of the origin and evolution of the solar system. A great array of observational facts must be explained by a satisfactory theory, and the theory must be consistent with the principles of dynamics and modern physics. All of the hypotheses so far presented have failed, or remain unproved, when physical theory is properly applied. The modern attack on the problem is less direct than the old method, which depended upon an all-embracing hypothesis. The new method is perhaps slower, but it is much more certain. By a direct study of the facts, we can specify, within an increasingly narrow range, the physical conditions under which the planets evolved. The manner of their origin must finally become apparent.

In the present chapter we will first coordinate some of the more basic observations related to the problem, review briefly the older hypotheses and their more obvious shortcomings, and then follow the first steps along the modern approach to the problem.

It is an interesting commentary on modern science that the *age* of the Earth has been determined, although its origin remains a baffling problem. The oldest rocks in the Earth's crust solidified more than 3 billion years ago and the materials of the Earth were brought together somewhat more than 4.6 billion years ago. Radioactive substances within the rocks leave minute traces of lead, helium, and other atoms to constitute a measure of the time elapsed since the Earth cooled. Studies of meteorites show that none of these visitors from space have been solid for measurably longer than the calculated age of the Earth. Since meteorites represent fragments of the solar system, we may conclude from their corresponding ages that the system is coeval with the Earth. The problem of the origin of the Earth is, therefore, synonymous with the problem of the origin of the entire system. Something happened nearly 5 billion years ago to generate the planetary bodies and to produce the order and regularity that we observe today.

Outstanding orderliness is conspicuous in the planetary motions. The members of the solar system move in the same direction along a common plane. Not only do the planets and thousands of asteroids follow this plane in their revolution about the Sun, but the great majority of the satellites move about their primaries in a similar fashion. The Sun, moreover, and 6 of the 9 planets exhibit the same phenomenon in their axial rotation. Even Saturn's rings share in the common motion. Of the few exceptions, we have mentioned the Uranus system, Venus, Neptune's Triton, and some of the outer satellites of Jupiter and Saturn; in addition, a fair fraction of the comets are included. We do not know the direction of Pluto's rotation.

The common motion of so many bodies suggests an initial rotary action, as though the solar system were once sent spinning by some cosmic finger. There is, in fact, so much motion in the outer bounds of the system that the older evolutionary hypotheses have failed in one respect; they cannot explain the *angular momentum* of the major planets. The angular momentum of a planet moving in a circular orbit at a given distance from the Sun (which is practically at the center of gravity of the solar system) is the product of its mass, distance, and speed. Since the speed diminishes only as the square root of the distance, a given mass contributes more angular momentum at a greater distance from the Sun. For a planet moving in an elliptical orbit, Kepler's law of areas (p. 24) expresses the constancy of the angular momentum at all times. When the planet is near the Sun it moves more rapidly than when it is farther away. No force toward or away from the Sun can change the angular momentum of a planet. Only an external push or drag along the orbit can increase or diminish this fundamental quantity of motion.

Jupiter, with its great mass, carries about six-tenths of the entire angular momentum of the solar system. The four giant planets contribute about 99 percent, and the terrestrial plancts 0.2 percent. The Sun, with a thousand times the mass of Jupiter, rotates so slowly that its angular momentum is only ½ percent of the whole. If all planets could be put into the Sun and could carry with them their present angular momentum, the augmented Sun would rotate in less than 10 hours, rather than a month.

A satisfactory hypothesis for the origin of the solar system must first account for the existence of the Sun, planets, satellites, asteroids, and comets. It must then explain how they were set moving in the remarkable manner already noted, and must provide the system theoretically with the observed amount of angular momentum. Two types of hypotheses have been suggested. In the first type, the system condensed from a gigantic cloud of gas and dust. In the second, the planets were derived from the gases of the Sun, either by an encounter with a passing star or by an ejection process. No form of either type of hypothesis has yet proved satisfactory, but both contribute greatly to astronomy by their impetus to thought.

The hypothesis that was believed for the longest time, excepting the Biblical account, was presented apologetically by the great French mathematician Pierre Simon Laplace (1749–1827), at the end of the eighteenth century; it was somewhat similar to an idea of the noted philosopher Immanuel Kant (1724–1804). According to this *nebular hypothesis,* a rotating and therefore flattened nebula of diffuse material cooled slowly and contracted. In the plane of motion, successive rings of matter were supposed to have split off, to condense into the planets of our present solar system. Most of the matter finally contracted to form the Sun. Between the present orbits of Mars and Jupiter, the ring failed to "jell," and produced many asteroids instead of a planet. The sequence of events is pictured in Fig. 172. Figure 173 shows a nebula—a spiral galaxy or island universe—from which the planets could certainly *not* have condensed.

Fig. 172. Laplace's nebular hypothesis. The condensation of a rotating gaseous nebula into the Sun, planets, and asteroids is here visualized. (Drawings by Scriven Bolton, F.R.A.S.)

The nebular hypothesis is untenable for several reasons, particularly because a speed of rotation sufficient to leave nebular rings at the present distances of the planets would provide the nucleus with many times the angular momentum of the rings. The Sun, according to the hypothesis, should have *more* angular momentum than the planets, not one-fiftieth as much. Furthermore, James Clerk Maxwell (1831–1879) showed that a fluid ring could not coalesce into large planets but would be transformed into a ring of *planetoids,* such as Saturn's ring or the belt of asteroids.

Collision or encounter theories attempt to avoid the difficulties of angular momentum. If another star collided with the Sun or passed very close to it, material might be ejected from its surface and condense to form the planets. Several variations of the encounter theory have been propounded. In the *planetesimal theory,* proposed early in this century by T. C. Chamberlin (1843–1928) and F. R. Moulton (1872–1952) of the University of Chicago, the passing star was supposed to have raised gigantic tides on the Sun. An appreciable quantity of matter, several times the present masses of the planets, was then ejected from the Sun's surface, and sent spiraling around it by the passing star. Most of the matter was lost or fell back into the Sun, but part remained, with a highly elliptical motion. The gases then condensed into small fragments, the

Fig. 173. The spiral nebula NGC 4736, photographed by the 200-inch reflector. Such a nebula contains more than a billion times the mass of the entire solar system, and therefore could not condense in the manner of Fig. 172. (Photograph by the Mount Wilson and Palomar Observatories.)

planetesimals, and as time progressed the larger fragments swept up the smaller, to form the planets. The rapid motion of the passing star provided the angular momentum for the orbital motions of the planets, their rotations, and the satellite systems. Within 20 million years after the encounter the formation of the planets would have been essentially complete.

Sir James Jeans (1877–1946) and Sir Harold Jeffreys proposed an alternative version of such an encounter. In their *tidal theory* a long tidal filament was drawn out of the Sun by the passing star. The inner part of the filament returned to the Sun, while the outer portion escaped into space. A central portion coalesced into a string of beadlike condensations, the embryo planets (see Fig. 174).

Jeffreys abandoned the tidal theory as untenable, and substituted a collisional hypothesis in which the approaching star brushed by the Sun in actual contact. The subsequent phenomena of the filament and planet formation follow essentially the same plan as in the original tidal theory. H. N. Russell (1877–1957) suggested that a companion star of the Sun was struck, and that the planets evolved from its debris. A number of serious objections have been

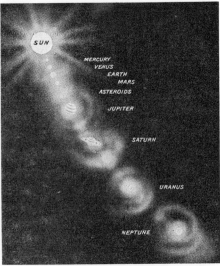

Fig. 174. The Sun encountered a star to produce the planetary system. An artist's concept of Sir James Jeans' tidal theory. (Drawings by Scriven Bolton, F.R.A.S.)

raised against all of these encounter hypotheses. In particular, when mathematical analysis is applied, the observed distribution of angular momentum in the solar system has not yet been explained. Furthermore, L. Spitzer has shown that in any tidal or encounter theory an appreciable fraction of a solar or stellar mass must be scooped up rapidly to form the planets. At the minimum required depth the temperature of the stellar material is a million degrees or more, normally held there by the great gravitational pressure of the overlying gas. Sudden removal to an essentially zero-gravity condition would cause an explosive expansion that would prohibit all condensation or contraction processes for producing planets or planetesimals.

Note, however, that the *planetesimal hypothesis* of Chamberlain and Moulton is not disproved by the arguments against the various tidal or encounter theories. Thus modern modifications of the Laplacian nebular hypothesis combined with the planetesimal hypothesis constitute in principle a valid approach to solar system evolution, and such an approach has been pursued by O. Schmidt of the U.S.S.R. and other modern students of the subject.

Clearly we are not yet developed enough scientifically to produce a complete and all-encompassing description of how the Sun and planets came into being. If we assume that the Sun developed with a large disk about it, we must find a means whereby the angular momentum was transferred from the Sun to the disk or to the materials that were lost from the system. C. F. von Weiszäcker in 1945 attempted to explain this momentum transfer by turbulent vortices from the Sun to the disk. His process must stop, however, when the Sun's equator rotates near orbital speed, more than ten times the present rate of rotation (Fig. 175). Furthermore, his idea that planets could form, like ball bearings, between two great vortices is quite unlikely, while material shot out from the Sun would fail to carry enough angular momentum to produce the planets whether the Sun were rotating rapidly or slowly.

Among stars like the Sun we find none that are rotating rapidly; also there is no other rational explanation for the formation of stars, of which the Sun is a typical example, except by collecting them from larger masses of gas and dust; the great clouds must inevitably contain a considerable amount of motion and therefore angular momentum. The author attempted in 1948 to avoid this problem by explaining the observationally unique solar system by

Fig. 175. Von Weiszäcker's eddies. Planets were presumed to form like ball bearings between the eddies and to rotate in the opposite direction. (From *Physics Today*, 1948.)

a theoretically rare process in which a condensing cloud of dust (and gas) *happened* not to carry much angular momentum. The condensing process in the *dust-cloud hypothesis* is also theoretically rare; the pressure of light from stars is assumed to act on the dust of interstellar space to produce a condensation. The suggestion appears to be unlikely to explain the solar system because of its rarity; also, stars produced in this fashion should generally be rotating very rapidly. The dust-cloud hypothesis, however, does contain one idea not much considered before, namely, that the preliminary concentration of material in the planets may have preceded the major concentration of material in the Sun.

Similarly W. H. McCrea has suggested that the great nebulous mass of material from which the solar system formed was composed originally of many separate blobs of gas and dust with a random motion of somewhat less than 1 mile per second. He then postulates that by chance those moving toward the same point in space coalesced with a minimum of net angular momentum to produce the partially condensed mass of gas from which the Sun and planets evolved. In this concept again one would expect most stars to be rotating near their limit of stability while the process appears highly unlikely.

In fact we find that the massive, hot, new stars are always rotating very rapidly, but the older, cooler, and smaller stars like the Sun are not. These facts highlight the existence of some process whereby the angular momentum of condensing stars is lost. Only one such mechanism has been suggested to retard the rotation of stars on such a universal scale. It involves the phenomena of *magnetohydrodynamics,* which concerns the effects of electric currents and magnetic fields in changing the physical laws for a hot conducting gas, and is important when the energy contained by the electromagnetic forces is comparable to the energy contained in the motions of the gaseous particles. In this physical state the lines of the magnetic field, maintained by electric currents in the ionized gas, add a kind of rigidity or wiry strength to the gas, which makes it resist distortion across the field lines by forces such as gravity, gas pressure, and differential rotation. Only magnetohydrodynamic forces in a hot gas *plasma* will ever make it possible to contain the enormous energies of nuclear fusion for the artificial transmutation of hydrogen into helium as a power source. H. Alfvén of Stockholm, Sweden was the first to apply the principles of magnetohydrodynamics systematically to astronomical problems.

Strong magnetic fields are present in the Sun, especially around sunspots. Radio noise from the Sun, and particularly the great outbursts of solar flares with associated cosmic rays and the solar wind, prove that in the Sun we are dealing with hot plasmas in which the embedded lines of magnetohydrodynamic force play a vital role. Young stars that will develop into the solar type show violent irregular variations of a similar nature. Since the late 1940's it has been clear that only magnetohydrodynamics presents a likely process for transferring the angular momentum away from a rapidly rotating new star to leave a slowly rotating star like the Sun.

The vital and unresolved question at the moment turns between two possibilities. Did the Sun transfer its angular momentum to a disk of gas already present in the region of the planets, thereby dissipating to infinity most of the light gases such as hydrogen and helium, and possibly increasing the angular momentum of the materials condensing into planetesimals and planets? Or were both the mass and the angular momentum carried out from the Sun, set spinning about it, and then partially condensed to produce the planetary system as we know it today? In 1960 F. Hoyle adopted a version of the latter concept in his theory of planetary evolution, the initial disk of the Sun extending to about the dimensions of Mercury's present orbit. In this concept the rapidly rotating Sun is connected by lines of magnetohydrodynamic force, analogous in some respects to long, elastic threads tied to the ionized material within the planetary disk (Fig. 176). Since the outer disk is turning more slowly, the threads tend to wind around and around and stretch, thereby increasing the angular momentum of the disk and slowing down the rotation of the Sun. Unfortunately Hoyle's mechanism fails in transferring planetesimals to the outer regions of the planetary system. The tight magnetic spiral produces an *outward* pressure on the gases as well as a small forward component along their direction of motion. Consequently the gas moves as though it were under less than the normal solar gravitational attraction and hence more slowly than the planetesimals. The planetesimals, therefore, meet an effective resisting medium and spiral *inward* while the gases spiral *outward*. The magnetohydrodynamic process of eliminating gas from the system and slowing the Sun's rotation is not challenged, but any mechanism that provides the total planetary mass from the inner solar region must be abandoned.

Let us now shift our attention to the chemistry and physics of the planets. This point of view was largely initiated by Henry Norris Russell, promoted by the thinking of Harrison Brown and others, and carried to a high level of confidence by H. C. Urey. First we note that the planets and asteroids within the orbit of Jupiter are made of solid rocks with a considerable amount of iron present and only a trace of the light gases such as hydrogen and helium, or even of the heavier *noble* gases such as neon, argon,

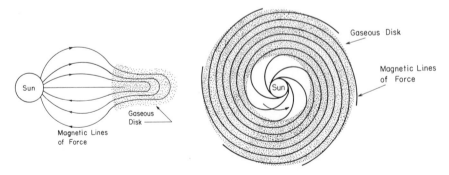

Fig. 176. Solar magnetic lines and the preplanetary disk, according to Hoyle: (*left*) solar magnetic lines of force connect across the plane of the ecliptic; the left-hand side of the "doughnut" is omitted; (*right*) view normal to ecliptic shows winding of solar magnetic lines in the material of the preplanetary cloud.

krypton, and xenon, which do not normally form compounds. On the other hand, the Sun and the stars that are close enough for spectral analysis consist almost entirely of hydrogen and helium. Perhaps 1 percent of the mass is contained in carbon, nitrogen, and oxygen together, while the remaining heavy elements constitute less than 0.5 percent, excepting neon which is probably more abundant than all the other elements heavier than helium. If, then, the Earth and other planets are to be made from material at all like the Sun in composition, what has happened to the light gases and to the noble gases? Looking to Jupiter and the other giant planets, we see that a large amount of hydrogen and helium are indeed present. Some of the satellites, too, appear to be made of light molecules such as possibly water, ammonia, and methane. Even Jupiter and Saturn, however, appear to contain a much larger fraction of the heavier elements than does the Sun, while Uranus and Neptune could be made solely of compounds, that is, with relatively little hydrogen and helium. This is exactly the composition of comets as deduced from the icy-comet model. We know too little about Pluto to consider it in this discussion.

Harrison Brown found that he could divide solar or stellar material into three basic classes, *earthy, icy,* and *gaseous* (see Table 3), based on striking differences in the melting (or freezing) points of the materials. The abundances are those for the Sun, assumed not to have changed greatly since the system was formed. All compounds are included in the earthy and icy groups, water (H_2O), ammonia (NH_3), and methane (CH_4) being ices, of course, at rather low temperatures. It is not certain whether hydrogen and helium will freeze at low pressures anywhere in the visible universe.

The division of the three kinds of material among the members of the solar system is summarized in Table 4. Near the Sun we find planets and asteroids consisting of materials that are solids at mod-

TABLE 3.   *Classes of solar materials.*

Classes	Elements	Mass abundance	Melting point
Earthy	Si, Mg, Fe, etc., plus 0*	1	~3600°F
Icy	C, N, O plus H*	4–7	≤0°F
Gaseous	H, He, Ne, etc.	300–600	≤14° absolute

* In chemical combination.

TABLE 4.   *Compositions in the solar system.*

Bodies	Earthy	Icy	Gaseous
Terrestrial Planets	0.7	0.3*	0
Asteroids	0.7	0.3*	0
Jupiter	<0.01	0.1	0.9
Saturn	0.01	0.3	0.7
Uranus	0.1	0.8	0.1
Neptune	0.2	0.7	0.1
Comets	0.15	0.85	0

* Oxygen, chemically combined in rocks.

erately high temperatures while at greater distances from the Sun the giant planets are made mostly of materials that boil at much lower temperatures. It is thus natural and reasonable to assume that *temperature* was the key factor in planetary composition, presumably controlled by solar distance. Near the Sun the higher temperatures must have permitted the heavier "earthy" materials to solidify without holding or gathering about them the huge fraction of gas that must originally have been present.

G. P. Kuiper, in his theory of the *protoplanets,* assumes that originally each planetary mass was a very large body spread over a great distance of the solar system with its original mixture of light and heavy elements. By applying the principles of gravitational stability set out in 1850 by E. Roche (1820–1883), Kuiper makes a case for the planets forming at the relative distances given by Bode's law (Appendix 1). In the original disk there was a greater concentration of material in the neighborhood of Jupiter and Saturn with the quantity falling off at greater and lesser distances from the Sun. The *proto-Earth* in this theory was once several hundred times more massive than the present Earth. The difficulty that has rendered Kuiper's theory practically unmanageable is that of removing the overwhelming fraction of hydrogen and helium from the huge protoplanets so that they could develop into planets, particularly into the terrestrial planets.

The more specialized chemical approach by Urey, on the other hand, has led to a widely accepted concept, namely, that the Earth and the asteroidal-meteoritic planets, on the basis of their present compositions, were formed at relatively low temperatures, below 2200°F, by the accretion of the solids of which they are now composed. Thus, within the present orbit of Jupiter, the temperatures must have been high enough in the forming system to keep water, ammonia, and methane as gases but low enough that earthy materials could condense into small solids. These planetesimals, then, accumulated into the terrestrial planets and into a number, perhaps large or small, of minor planets which collided to produce the present asteroids and the meteorites that fall on the Earth. Thus Urey found chemically a firm basis for the planetesimal hypothesis, which was a major part of the earlier theory by Chamberlin and Moulton. The comets can now play the role of planetesimals for the outer planetary system, Uranus and Neptune (and Pluto?) having much the composition expected of large aggregates of comets.

Thus we have a gross, fairly reliable picture, without details, of a process in which, near the Sun, earthy materials condensed and collected into larger and larger aggregates until they made the terrestrial planets. This picture is not unlike that developed independently by O. Schmidt (1891–1956). The gases, including the noble gases and hydrogen, carbon, nitrogen, and oxygen, that must have been present at the time of the original condensation or collection were in some fashion removed so that they did not accumulate on the larger masses as the collection process culminated in the present terrestrial planets and asteroids. At the median distance of Jupiter and Saturn the loss of gases was less extensive.

Continuing, then, on the hypothesis that the present solar mix represents the original distribution of elements in the solar-planetary nebula, we can calculate the *minimum* mass required to make the planets. Table 5 lists the present masses of solar-system objects, the multiplying factors to provide the present masses from a solar mix, and the minimum masses needed to form the objects by a 100-percent efficient process, discarding the unused ices, gases,

TABLE 5.   *Present solar-system masses and original minimum masses.*

Bodies	Present mass (Earth = 1)	Factor	Original minimum mass (Sun = 1)
Terrestrial	2	400	0.002
Jupiter	318	11	0.011
Saturn	95	36	0.010
Uranus and Neptune	32	80	0.008
Comets	1	8,000?	0.024?
Total minimum original mass			0.055?

or both. We see that 6 percent of a solar mass would be adequate, but since the efficiency must have been very low, perhaps an entire solar mass was actually involved originally. Note that the quantity of solar material needed to yield each type of object is comparable.

For the comets I assume that the efficiency of formation was a hundred times lower than for the remainder of the planetary system. G. P. Kuiper, A. G. W. Cameron, and I visualize the comets as being formed beyond Saturn in a large belt near the plane of the planets. These *cometesimals* collected to form Uranus and Neptune, and an appreciable number were captured by Saturn to give it more icy and earthy content than Jupiter. Both Jupiter and Saturn must have formed quickly, the original mass of Saturn being appreciably less than its present mass.

As Uranus and Neptune gained mass, they began to perturb the comet orbits gravitationally at close approaches. Hence many comets were thrown into the inner system where they were vaporized by solar heat as comets are today. Many were captured by the Sun and planets, while others were perturbed into hyperbolic orbits about the Sun, to be lost forever from the solar system. The remainder, perhaps on the order of 1 percent, were thrown into long-period orbits to constitute the cloud of comets about the Sun, extending out to some 100,000 A.U. This cloud of comets was shown by E. Öpik to be stable against disruption by passing stars for the life of the solar system, since each passing star would only eliminate a small fraction by perturbations. He has also shown that originally the comets could be thrown into such very large elongated orbits by the perturbations of Uranus and Neptune. J. Öort found that the passing stars could perturb comets from the cloud into orbits near the Sun and thus maintain a supply of new comets at the present time (see *Between the Planets* by F. G. Watson, Harvard University Press, Cambridge, Mass., 1956). Since Öort estimates the present total mass of the comet cloud at something like an Earth mass, the original mass of the comets must have been much larger; hence an additional factor of 100 (100×100) for losses in Table 5.

If cometesimals formed Uranus, Neptune, and the present comet cloud, a large number of small ones must still exist in a *comet belt* beyond Neptune (see Fig. 177). The existence of such a belt has not been demonstrated because comets would be unobservable from the Earth at such distances. Also the integrated mass would be small and not produce major perturbations. Eventually, however,

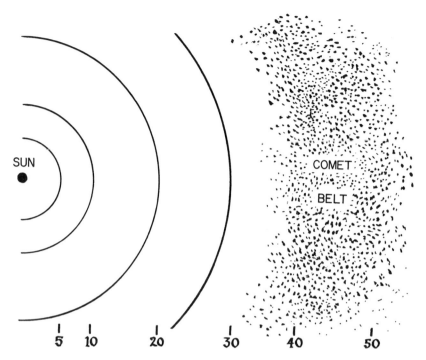

SUN

COMET

BELT

| | | | | | |
| 5 | 10 | 20 | 30 | 40 | 50 |

Fig. 177. Region of possible comet belt. Near plane of the ecliptic beyond 30 astronomical units distance from the Sun.

the comet belt may be detected by perturbutions in the motions of long-period comets. The lack of conspicuous effects in the 1910 return of Halley's Comet has led B. Marsden, S. Hamid, and the author to place the maximum mass to a distance of 50 A.U. and as less than one Earth mass. The author's suggestion that the comet belt produced the perturbations in Neptune's motions once ascribed to Pluto's attraction would require several Earth masses in the present comet belt and cannot now be accepted as likely.

That large clouds of gas and dust can collapse to form stars is evident from the identification of a few such new stars among the T Tauri variable stars, particularly by the work of G. Herbig. The T Tauri stars occur in large Milky Way clouds of gas and dust and show clearly that they are ejecting matter at a very rapid rate. C. Hayashi has demonstrated that the collapse of a stellar mass from an extended cloud can occur very rapidly, almost at the rate of free fall. The energy of collapse goes into dissociating the hydrogen molecules and then into ionizing them and helium, the two elements constituting almost all the mass. After the major collapse to a radius about the size of Mercury's present orbit, for some hundreds of thousands of years the star becomes several magnitudes brighter than its final magnitude will be when it settles down to a semi-stable system of energy generation such as operates in the Sun today. Remarkable discoveries of infrared stars by E. E. Becklin and by D. E. Kleinmann and F. J. Low demonstrate unquestionly the present-day occurrence of proto-solar systems in dust clouds. The detailed theory of such collapse and the prediction of the resulting circumstances in the nebular rings or inner regions is extremely difficult, so that no detailed picture of the development of a Laplacian nebula can be accepted as yet. Even the conditions that produce a planetary system rather than a double star are difficult to define.

The observed new stars occur in regions where other stars are forming or have formed recently; exploding massive stars or *supernovae* leave traces in the abundance ratios of the elements. The study of meteorites, amazingly enough, leads to direct evidence that such was probably the case for the solar system. J. Reynolds found that the isotope xenon-129 occurs with unusually high abundance as compared to other xenon isotopes in certain meteorites. The only likely parent atom of xenon-129 is iodine-129, which is radioactive with a half-life of 17.2 million years. Since the xenon, a noble gas, must have been formed and held in the asteroidal body itself, the iodine-129 atoms must, therefore, have been created not too many tens of millions of years prior to the formation of the asteroids. Supernova explosions could create such atoms. Thus some of the material of prior supernovae was actually incorporated into the primordial matter of the solar system.

Why little asteroids rather than a sizable planet were formed between Mars and Jupiter remains unexplained. Probably Jupiter is responsible. Forming early in the nebula, the great mass of Jupiter may have disturbed the motions of the nearby gas, dust, and planetesimals so that accretion processes were slowed down. Once the "solar gale" of matter thrown out by the Sun began to dissipate the hydrogen and helium of the solar nebula, the many asteroids began to disintegrate slowly by collisions, as they do today.

At about this time something drastic happened to Earth's primitive atmosphere, accumulated from the solar nebula. The extreme rarity of the heavy noble gases in our atmosphere today and the calculated water sources from volcanic action indicate that the original atmosphere and most or all of the water was lost from the earth. Is it possible that the Earth at one time was subjected to an enormous "bath of fire" which might conceivably have been extensive enough or prolonged enough to boil away early oceans and atmosphere?

Some time after the Earth's major accumulation, its interior was heated by radioactivity, so that the dense iron-nickel could settle to form the core and the silicate outer mantle could develop to its present state. Almost all of the radioactivity was carried chemically to the mantle; we observe the same phenomenon in the meteorites, the irons containing almost no radioactive uranium or thorium. We note then that Mars appears, from its mean density, to have a relatively smaller iron core than the Earth while Venus is not greatly different. Mercury must contain a much greater proportion of heavy elements than any of the other terrestrial planets. Does that sequence in average composition mean that iron froze out more rapidly near the Sun than the silicate compounds? Or could the "bath of fire" have removed a thick layer of silicates from the surface, more from Mercury than from the more distant planets? Whether or not the Sun actually expanded for a short time to envelop the inner planets is unlikely, but such an event might account for these additional facts concerning the terrestrial planets. The low density of the Moon (Chapter 9), however, presents a serious problem.

Three possible mechanisms for producing the Moon are seriously discussed today. The most obvious is that many "moonlets" collected in a ring near the Earth. The tidal friction caused a larger and inner moonlet to spiral outward, growing as it spiraled, and swallowing up all of the outer ones. The remaining inner ones, being small, were dissipated away, perhaps by the violence that removed the Earth's original atmosphere, perhaps by perturbations or perhaps by collisions with debris. This "obvious" method of making the Moon gives no clue as to its very low mean density.

A second possibility is that the Moon was captured, ready made. A mechanism of capture, developed by H. Gerstenkorn, is not impossible. But why should the Moon's density be lower than that of the other bodies formed in this region of the solar system?

A third possibility is a modern version of Sir George Darwin's tidal separation of the Moon from the Earth. Suppose, with D. U. Wise, that when the Earth formed it was rotating very rapidly, near the limit of stability. As radioactivity melted the interior, the dense iron settled to form the core. The total angular momentum, of course, remained constant but the mantle became less dense and contained a smaller fraction of the total mass. As a consequence the Earth turned faster. Thus the atmosphere and the outer mantle near the equator were thrown out to maintain stability. Probably in such circumstances, the Earth would have elongated something like a tenpin and suffered fission by throwing out the Moon mass from the small end. Tidal friction then moved the Moon to its present position. This fission theory of the Moon is highly attractive because it forms the Moon out of the lower density mantle of the Earth and simultaneously removes the Earth's primitive atmosphere. The theory, however, lies on the verge of the impossible because of the limited effects arising from the settling of the iron core.

The detailed formation of the other satellites in the solar system remains a subject fraught with great uncertainty. In Kuiper's theory, which now seems largely untenable for the terrestrial planets, satellites were formed in great numbers within their huge protoplanets somewhat as the planets formed about the Sun. As the protoplanets lost their masses many of the satellites escaped; from Jupiter some became the *Trojan asteroids,* those that move in essentially the same orbit as Jupiter, positioned around the orbit about 60° on either side of the planet. If, on the other hand, planetesimal accumulation developed the major planets, some satellites may have been captured as the planets grew. We need to account for the fact that the outer Jovian satellites appear to be of low density and the inner ones more rocky, like the Moon.

Many problems of solar system evolution remain unmentioned or untreated in this brief review of the subject, but research progress is extremely rapid. Direct space exploration coupled with the unbelievable sensitivity of chemical and isotopic analysis is providing tools of such power that clear-cut answers to many old problems are in sight. Furthermore the observations of infant stars in stellar nurseries give us both confidence and insight concerning hitherto baffling aspects of planetary evolution.

The question as to the numbers of planets about other stars remains unresolved. Current opinion strongly favors the concept that almost all stars form in essentially the same fashion. Only in one respect is the Sun atypical; it is a single star having planets. A majority of stars appear to be double. Almost certainly most double stars cannot hold stable systems of nearby planets, although some could do so. It is not at all unlikely that one star in a hundred may be single and thus *able* to possess planets. In our own Galaxy of

10^{11} stars, it is not unlikely that one out of a million to perhaps one out of a thousand stars may have a planet like the Earth moving under conditions that are comparable to the solar radiation on the Earth. This leaves us in our galaxy perhaps a million to a billion planets on which sentient beings might develop; and there are millions of galaxies.

There is little doubt but what life forms develop where circumstances are favorable. The time of development for sentient beings, however, is extremely long, at least as evidenced in our fossil records. Man has been on this planet for less than 0.1 percent of its existence. If he can persist for 100 million years, 2 percent of the present age of the Earth, many skeptics will have erred. If so, however, there is a chance that perhaps 10,000 to 10,000,000 planets in our own Galaxy now contain sentient beings more or less like ourselves. Space travel to visit them is now virtually impossible in view of the limitation of travel imposed not only by the Einstein theory of relativity but by *laboratory experience*. We and our machines are made solely of atomic nuclei and electrons. In high-energy accelerators both types of particles are given enormous energies that would be sufficient under Newtonian theory to set them moving at huge multiples of the velocity of light. The multiple is about 100 times for protons and electrons, millions in cosmic rays. The atomic nuclei and electrons cannot be made to travel faster than light, however. As their energy increases they approach this critical velocity while their mass increases precisely according to relativity theory. Our high-energy accelerators would fail to operate if this were not true. Since we cannot make our component parts move faster than light it is obviously impossible for us to make our bodies do so, or our machines.

As for the same limitation on signals, however, a new possibility has been suggested by G. Feinberg. He theorizes that there may be particles, *tachyons*, that cannot move *slower* than light. His postulated tachyons lose energy as they speed up. Thus relativity theory may conceivably allow signals to be sent at hyper-light velocities, even though ordinary matter will always be chained. If tachyons can first be demonstrated and then controlled, they present a possibility for communication with other intelligent beings in the universe. The only other possibility for such communication within the foreseeable future is the remote chance that such beings might wish to signal us by radio or the greater chance that they might use such powerful transmitters in their own communications that our great radio telescopes could intercept their signals. A serious attempt to listen for such signals has been conducted by the National Radio Astronomy Observatory at Green Bank, West Virginia, under the code name Project OZMA. The arguments for and against the allocation of great effort in such a program are thought provoking. Much as the author likes the imaginative, science-fiction aspects of the program, he must express some skepticism regarding its success in the near future.

As for interstellar travel, the limitations both on energy and on velocity make the practical round trip to the nearest star, 4.3 light years away (see Table 7, p. 000), an operation that would span generations. Besides, with today's technology, it might create a budgetary catastrophe. Such a venture is not impossible but it is fantastically improbable until more powerful and much cheaper power sources become available. Fusion power, however, may be just beyond today's horizon. Success in detecting extraterrestrial cultures or, better, in communicating with them would, of course, crystalize world interest and might trigger a colossal international effort to explore interstellar space. Nothing, at least in this writer's view, could be more exciting or more stimulating to the human race.

We have studied the present state of our planetary system, and, to a limited extent, its history. Its future, unless some unforeseen accident occurs, seems bright. The chance that a wandering star might disrupt the stately order of the planetary motions is small, even within a billion years. Nor should we expect a great change in the Sun's radiance much sooner than this. Probably the glacial ages will recur; we cannot say. The continents may rise and fall during the ensuing ages, as they have done in the past—we hope they do it slowly. And random meteoric masses may pierce the surface here and there.

But order, which is the solar system, will prevail.

Bode's Law

The so-called Bode's law, ascribed to J. E. Bode (1747–1826), is not a physical law but only a convenient rule for recalling the distances of the planets from the Sun. Write down a series of 4's, one for each planet. Add to the successive 4's the numbers 0 for Mercury, 3 for Venus, 6 for the Earth, 12 for Mars, 24 for the asteroids, and so on. Insert a decimal point in each sum to divide by ten. The resultant series of numbers represents approximately the distances of the planets from the Sun, in astronomical units. The scheme of numbers follows:

	Mer.	Ven.	E.	Mars	Ast.	Jup.	Sat.	Ur.	Nep.	Pl.
	4	4	4	4	4	4	4	4	—	4
	0	3	6	12	24	48	96	192	—	384
Bode's law	0.4	0.7	1.0	1.6	2.8	5.2	10.0	19.6	—	38.8
Actual	0.39	0.72	1.00	1.52	—	5.20	9.54	19.18	30.07	39.67

Note that Bode's law includes the asteroids and gives the distance for Pluto rather than Neptune. The law was used in Leverrier's and Adams' predictions of the position of Neptune. The predicted orbits were therefore considerably in error.

No theoretical basis for the rule has been generally accepted.

Planetary Configurations

The various geometric positions of the planets with respect to the Sun and the Earth are known as the *planetary configurations*. They are shown in Fig. 178. For an observer on the Earth the angle between a planet

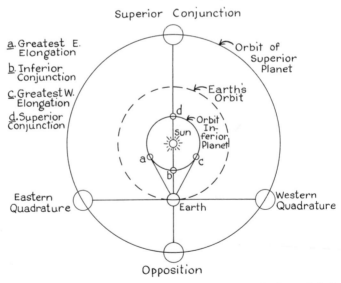

Fig. 178. Planetary configurations as seen from the Earth, for an inferior and a superior planet.

and the Sun is the *elongation*. A *superior* planet, whose orbit lies outside the Earth's, passes through all elongations to 180° east or west. An *inferior* planet, whose orbit lies within the Earth's, may attain only a certain maximum elongation: the *greatest eastern elongation* or *greatest western elongation*. An inferior planet comes to *inferior conjunction* when passing the line between the Earth and the Sun, and to *superior conjunction* when in line with the Sun beyond it.

A superior planet also may come to *superior conjunction* (or simply to *conjunction*). When directly opposite to the Sun, the configuration is *opposition*. The configurations at right angles to the Sun's direction are *eastern* or *western quadrature*.

An inferior planet is best observed near greatest elongation, east or west. A superior planet is best observed at opposition.

Planetary and Other Data

The following comments on the data of Table 6 are numbered to correspond to the number of the line.

1. One astronomical unit is the mean distance from the Earth to Sun. It is 92,956,000 miles (149,598,000 kilometers). The mean distance from the Earth to the Moon is 238,856 miles (384,402 kilometers); maximum, 252,710 miles; minimum, 221,463 miles.

2. The sidereal period is the time of one revolution with respect to the stars. One tropical year (of the seasons) is 365 days 5 hours 48 minutes 46.0 seconds. It is the unit of lines 2 and 3.

3. The synodic period is the time of one revolution with respect to the Sun as seen from the Earth.

4. For definition see page 23.

6 and 7. Mean values are given. The Earth's equatorial diameter is 7926.4 miles (12,756.3 kilometers) and its polar diameter is 7899.8 miles (12,713.6 kilometers).

10. The Earth weighs 6,600,000,000,000,000,000,000 tons (5.976 × 10^{21} metric tons).

12. The weight of a given object on the Earth when multiplied by the quantity in Table 6 becomes the weight of the object at the surface of the planet.

13. An object shot away from the equator with this velocity would escape forever into space (neglecting friction with an atmosphere).

14. The Sun's period of rotation varies from 25.0 days at its equator to 26.6 days at latitude 35°.

16. CO_2 is carbon dioxide. See Table 1 for the composition of the Earth's atmosphere. CH_4 is methane, NH_3 is ammonia, HCl is hydrochloric acid, and HF is hydrogen fluoride. The clouds of the giant planets may consist largely of ammonia crystals. Helium is probably very abundant in their atmospheres.

18. The *albedo* is the ratio of the total amount of light reflected by the planet to the light incident on it.

19. The inclination of the equator is given with respect to the orbit plane of the planet (Moon and Sun to ecliptic).

Miscellaneous Data

Velocity of light: 186,282.1 mi/sec = 299,792.5 km/sec

Constant of gravity in Newton's law, Force = Gm_1m_2/r^2 : $G = 6.67 \times 10^{-8}$ cm³/(gm sec²)

Parallax of the Sun: 8.7942 sec arc.

Time required for light to travel 1 astronomical unit: 499.005 sec

Light year, the distance light travels in one year: 5,878,500,000,000 mi = 9.4605×10^{17} cm

Surface gravity of the Earth at the equator: 978.036 cm/sec²

Rotational velocity of Earth's equator: 1035 mi/hr = 0.4626 km/sec

1 meter = 39.37 inches precisely

1 kilometer (km) = 0.621370 mile; 1 mile = 1.60935 km

1 kilogram (kg) = 1000 grams = 2.20462 pounds (lb)

1 metric ton = 1000 kg = 2204.62 lb = 1.10232 tons (ordinary)

Temperature in degrees Centigrade (°C) = ⅝ (Temp. °F − 32°)

Temperature in degrees Fahrenheit (°F) = ⅝ (Temp. °C) + 32°

Table 6. *Planetary data.*

No.	Datum	Mercury	Venus	Earth	Moon	Mars	Jupiter	Saturn	Uranus	Neptune	Pluto	Sun
1	Mean distance to Sun	0.387	0.723	1.000	1.000	1.524	5.203	9.540	19.18	30.07	39.67	0
2	Sidereal period	87^d97	224^d70	365^d256	27^d32	687^d0	11^y86	29^y46	84^y01	164^y8	249^y9	—
3	Synodic period	115^d88	583^d92	—	29^d53	779^d9	1^y092	1^y035	1^y012	1^y006	1^y004	—
4	Eccentricity of orbit	0.206	0.007	0.017	0.05	0.093	0.048	0.056	0.047	0.009	0.247	—
5	Inclination of orbit to ecliptic	$7°0$	$3°4$	$0°0$	$5°1$	$1°8$	$1°3$	$2°5$	$0°8$	$1°8$	$17°2$	—
6	Orbital velocity (mi/sec)	29.8	21.8	18.5	0.64	15.0	8.1	6.0	4.2	3.4	3.0	28
7	Equatorial diameter (mi)	3025	7526	7926.4	2160.6	4200	88,700	75,000	29,600	27,600	3600?	865,000
8	Polar flattening		?	1/298.35	1/2000	1/120?	1/16	1/10.2	1/17	?	?	—
9	Volume (Earth = 1.0)	0.056	0.857	1.000	0.0203	0.150	1318.	769.	50.	42.	0.1	1,304,000
10	Mass (Earth = 1.0)	0.055	0.816	1.000	0.01229	0.1070	317.9	95.2	14.5	17.4	?	332,950
11	Density (water = 1.0)	5.5	5.25	5.52	3.34	3.96	1.33	0.68	1.60	2.3	?	1.41
12	Surface gravity (Earth = 1.0)	0.38	0.89	1.0	0.165	0.38	2.6	1.1	0.96	1.5	?	28
13	Velocity of escape (mi/sec)	2.6	6.4	6.95	1.475	3.1	37.	22.	14	15	?	383
14	Period of rotation	59^d	243^d	23^h56^m	27^d3	24^h6	9^h8	10^h2–10^h6	10^h8	15^h7	153^h	25^d
15	Maximum surface temperature (deg. F)	600°?	600°?	140°	212°	75°	−227°?	−290°?	−300°?	−320°?	−360°??	10,000°
16	Gases identified in atmosphere	None	CO_2, H_2O? HCl, HF	Many	None	CO_2, H_2O	CH_4, NH_3 H_2, He?	CH_4, NH_3 H_2, He?	CH_4, H_2	CH_4, H_2	None	Many
17	Number of satellites	0	0	1	0	2	12	10?	5	2	0	—
18	Albedo	0.059	0.85	0.35	0.07	0.15	0.58	0.57	0.80	0.71	0.15?	—
19	Inclination of equator	<30°	177°	$23°5$		$24°0$	$3°1$	$26°7$	$97°9$	$28°8$	?	$7°2$

The Star Chart

The accompanying star chart (Fig. 179) is intended primarily for use with Table 8 (Appendix 5) to locate and identify the planets at any time from 1966 through 1980. A chart of the north polar region is shown in Fig. 180.

The *magnitudes* of stars define their brightness on a reversed scale. A first-magnitude star has the average brightness of the 20 brightest stars in the sky. A sixth-magnitude star is just one-hundredth as bright, and can barely be seen with the naked eye on a very clear dark night. Each magnitude denotes a step of 2.512 times ($\sqrt[5]{100}$) in brightness. Thus a star of the sixth magnitude is 2.512 times as bright as one of the seventh magnitude, and a hundred times as bright as one of the eleventh magnitude.

For the most brilliant stars the values become negative. Sirius, the brightest star in the entire sky, is of magnitude -1.42. Venus, the brightest planet, sometimes reaches a magnitude of -4.3. It is then more than a hundred times brighter than a first-magnitude star. Jupiter, at maximum, reaches a magnitude of -2.5, Mars -2.8, Saturn -0.4, and Mercury -1.2. Uranus is of magnitude 5.7, theoretically visible to the naked eye, but seen by very few individuals. Neptune is

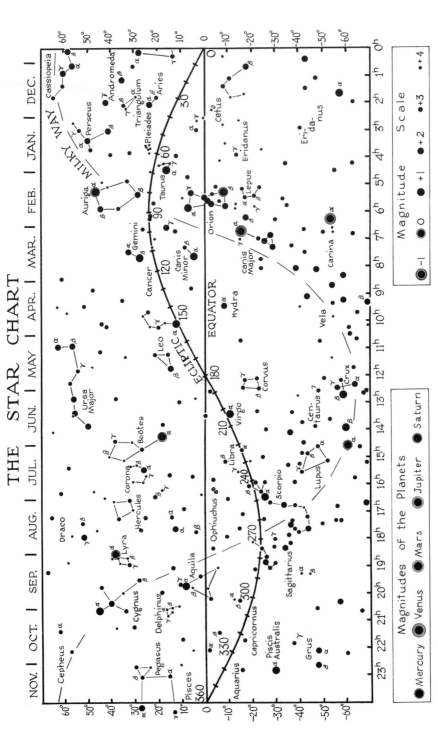

Fig. 179. Star chart, for use in conjunction with the planetary tables of Appendix 5.

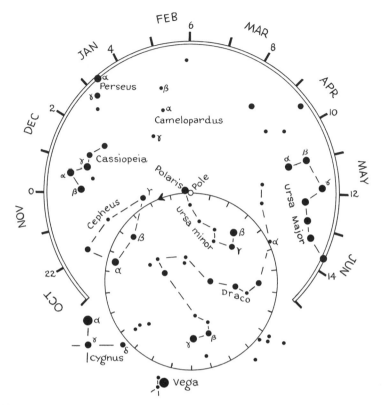

Fig. 180. Chart of the north polar constellations. Turn the chart so the correct month is uppermost and face north for observations near 8 P.M. The smaller circle shows the path of the pole of the ecliptic due to precession of the equinoxes. Each mark on the circle represents 1000 years and the whole motion is completed in about 26,000 years. In A.D. 14,000 Vega will be the pole star. (Chart by Donald A. MacRae.)

of magnitude 7.6, easily visible in quite a small telescope, but more difficult to detect than Uranus.

Venus, Jupiter, and Mars are brighter than any star, hence easy to identify in the sky. Saturn is also easily visible, as only a few stars exceed it in brightness. The planets may sometimes be identified by the steadiness of their light; the stars twinkle much more violently. Mercury is always so near to the Sun that no attempt should be made to find it except at its greatest elongation, and then only under favorable observing conditions.

The constellation identification for groups of stars arose in ancient times; the names are mostly Latin, Greek, and Arabic. In recent years the boundaries of the various constellations have been assigned by international agreement. The brightest star in a constellation is generally called alpha (α), the second beta (β), the third gamma (γ), and so on through the Greek alphabet, although there are many exceptions. (The possessive case of the constellation name is used in such designations.) Many of the stars have proper names a well. The most conspicuous constellations are labeled on the chart and the stars α, β, and γ are designated in many of them. The chart is complete to the third magnitude (with only two or three exceptions) and shows a number of the fourth-magnitude stars.

The names of some of the brightest stars are given in Table 7.

The horizontal line across the middle of the chart is the *celestial equator* on the sky. In the United States it can be located when one faces south and looks up at an angle of from 40° to 65° (90° minus the observer's latitude).

The long curved line that crosses the celestial equator is the *celestial ecliptic*. The planets all appear within about 7° of the ecliptic. The broad band, with the ecliptic as its central line, that includes the paths of the planets is called the *zodiac*. The numbers along the ecliptic give, in degrees, the *celestial longitude,* corresponding on the sky to ordinary

TABLE 7.   *The brightest stars.*

Name		Visual magnitude	Distance (light years)
Constellation	Common		
α Canis Majoris	Sirius	-1^m42	9
α Carinae	Canopus	-0.72	100?
α Centauri*	—	-0.25	4.3
α Bootis	Arcturus	-0.06	36
α Lyrae	Vega	0.04	26
α Aurigae	Capella	0.05	47
β Orionis	Rigel	0.14	800?
α Canis Minoris	Procyon	0.38	11
α Orionis	Betelgeuse	(0.41)	500?
α Eridani	Achernar	0.51	70?
β Centauri	—	0.63	300?
α Aquilae	Altair	0.77	16
α Tauri	Aldebaran	0.86	53
α Virginis	Spica	(0.91)	260?
α Scorpii	Antares	(0.92)	400?
β Geminorum	Pollux	1.16	40
α Piscis Australis	Fomalhaut	1.19	23
α Cygni	Deneb	1.26	1400?
α Leonis	Regulus	1.36	75
α Crucis	—	1.39	400
α Geminorum	Castor	$+1.97$	45

Data compiled by G. H. Herbig and C. E. Worley. The light year is defined on page 267.

* Its fainter companion, Proxima Centauri, is the nearest star.

longitude on the Earth. (Table 8 gives the celestial longitudes of the planets at convenient intervals of time from 1966 through 1980.) The scale along the bottom of the chart is the *right ascension*, measured in hours of time; 24 hours = 360°, therefore 1 hour = 15°. Right ascension is like longitude except that it is measured along the celestial equator instead of along the ecliptic.

The vertical scale at the sides of the chart gives the *declination*. It is measured north (plus) and south (minus) from the equator. Declination on the sky is precisely analogous to latitude on the Earth.

The average plane of the Milky Way, or Galaxy, is indicated.

The star chart was drawn by Donald A. MacRae.

To Use the Star Chart

Face south with the chart before you.

If the time is 8 P.M., the constellations under the current month will appear in the South. The map will extend over your head and below the horizon.

At 10 P.M., look under the next month, at 12 P.M. under the second month, and so on. At 6 P.M., look under the preceding month.

If you are located in the Southern Hemisphere, hold the chart upside down and face North.

The Planet Finder

Table 8 gives the positions of the brighter planets from 1966 through 1980. It was especially calculated for this volume from the tables by Karl Schoch and from the tables by William D. Stahlman and Owen Gingerich (Solar and Planetary Longitudes for Years −2500 to +2000, Madison, Wis., 1963), courtesy The University of Wisconsin Press. It is designed for use with the accompanying star chart (Fig. 179). The numbers given are the celestial longitudes in degrees. The planets can be located near the ecliptic on the star chart at the corresponding values of the longitude. *Italics* are used for the longitudes of the *morning* sky.

The Sun's longitude is given for the evening of the 13th of each month. The superior planets, Mars, Jupiter, and Saturn may be difficult to find when they are near to the Sun on the sky, that is, near conjunction; the date immediately preceding conjunction is indicated by a dagger †. They can be observed during the entire night when near opposition, indicated by an asterisk *. They are visible in the evening sky for about 2 months after opposition. Jupiter always lies within 2° of the ecliptic, Saturn 3°, and Mars 7°. For Mars this large deviation occurs only when the planet is near opposition; see Fig. 153 for the most favorable oppositions. Mercury moves so very rapidly that is longitudes are given at 10-day intervals. When no longitude is given in Table 8, Mercury is hopelessly close to the Sun. Mercury's distance from the Sun at greatest elongation varies enormously because of the eccentricity of its orbit. Each greatest elongation is indicated by at least one value of the longitude. When values of the longitude are given for four consecutive 10-day intervals in Table 8, there is a fair chance, with the naked eye, to identify Mercury on the sky. Choose a night near the middle of the series. When only one or two consecutive values are given there is little chance of finding Mercury without binoculars. When looking for Mercury it is best to estimate the longitude at the date between the values given. For the longitudes given, Mercury will always lie within 6° of the ecliptic.

The longitudes of Venus are given at 15-day intervals. The planet is so bright that it may occasionally be seen very close to the Sun at dates for which the longitude was not calculated, but at such times it will be visible only for a short time in the evening or morning twilight. Venus may lie as much as 7° from the ecliptic.

Venus and Mercury can be seen in the evening when their longitudes are not in italics. They can be observed in the morning sky only when the longitudes are in italics.

If instead of the star chart in Fig. 179 you wish to use some other chart, divide the numbers in Table 8 by 15, to convert them into hours of time. Then use the Right Ascension scale of the other chart and locate the planets near the ecliptic as with the present chart.

How to Find the Planets

Locate the year and the month in Table 8. Read the number in the table for each planet at the nearest date. Locate this number along the curved ecliptic on the Star Chart. Look for the planet near the ecliptic at this point. It is well also to locate the position of the Sun.

If no number is given in Table 8, the planet is too close to the Sun for observation. Do not look for a planet when the number carries a dagger † or immediately follows such a number. Note the italics (000) for planets in the morning sky. For about two months before an * in the table, a planet can be seen in the east in the evening.

Venus and Jupiter are brighter than any star. Mars may be a bit fainter than Sirius, the brightest star. Saturn is somewhat fainter but still bright. Mercury is very difficult to find.

Read Appendix 2 for definitions of planetary configurations and Appendix 4 for a description of the star chart.

Example: To locate the planets on Christmas Eve, December 24, 1976. On page 279, for December 1976 we find that the entries are 262° for the Sun on December 13, 291° for Mercury on December 23, 322° for Venus on December 28, *256°* for Mars, 52° for Jupiter, and *137°* for Saturn. Thus Jupiter, near opposition, is ideally suited for late evening observation and is near the Pleiades. Venus is conveniently located in the evening sky but rather far south for observation from the northern hemisphere. Mercury will be *very* difficult in the evening as will Mars in the morning, and Saturn will be a morning object.

TABLE 8. *The Planet Finder.*

Planet	Sun	Mercury			Venus		Mars	Jupiter	Saturn
Date	13th	3rd	13th	23rd	13th	28th	13th	13th	13th
1966									
January	294	—	—	—	—	—	317	83	343
February	325	—	—	—	299	302	342	81	346†
March	353	1	—	—	310	322	4	82	350
April	24	352	357	6	337	353	27†	86	354
May	53	20	—	—	10	27	49	92	357
June	82	—	—	117	45	63	71	98†	359
July	111	127	132	—	81	99	91	105	359
August	141	—	123	—	118	—	113	112	358
September	171	—	—	—	—	—	132	118	356*
October	200	—	221	234	—	—	151	122	353
November	231	242	—	—	—	—	169	124	353
December	262	231	—	—	—	—	185	124	353
1967									
January	293	—	—	—	—	—	200	120*	355
February	325	—	342	—	348	7	210	116	358
March	353	—	—	337	22	41	213	115	1†
April	23	346	9	—	60	77	205*	116	5
May	52	—	—	—	94	111	196	119	8
June	82	95	106	112	128	142	197	124	11
July	111	—	—	—	154	162	208	130†	12
August	141	112	—	—	163	—	223	137	12
September	170	—	—	202	—	149	242	144	10
October	200	215	225	—	157	169	263	149	8*
November	231	—	212	—	184	201	286	154	6
December	262	—	—	—	218	236	309	156	5
1968									
January	293	—	—	—	255	273	334	155	6
February	324	332	—	—	292	311	358	152*	9
March	354	318	326	338	328	346	20	148	12†
April	24	—	—	—	—	—	42	146	16
May	53	—	—	86	—	—	64	147	20
June	83	—	—	—	—	—	85†	150	23
July	112	—	91	—	—	—	105	155	25
August	141	—	—	—	—	—	125	161†	25
September	171	183	197	207	194	213	145	168	24
October	201	211	—	—	231	250	164	174	22*
November	232	204	—	—	269	287	183	180	20
December	262	—	—	—	305	322	201	184	18

000 *Evening.* 000 *Morning.* † *Conjunction with Sun.* * *Opposition to Sun.*

TABLE 8 (continued)

Planet	Sun	Mercury			Venus		Mars	Jupiter	Saturn
Date	13th	3rd	13th	23rd	13th	28th	13th	13th	13th
1969									
January	294	—	313	—	340	356	218	186	19
February	325	—	301	309	11	22	235	185	21
March	353	318	342	—	27	—	247	182*	24
April	24	—	—	—	—	10	256	178	28†
May	53	64	—	—	14	24	255	176	31
June	83	—	—	70	37	52	245*	177	35
July	111	—	—	—	68	84	242	180	37
August	141	—	162	176	103	120	250	185	39
September	171	188	195	—	139	157	266	191†	38
October	200	—	182	—	176	194	285	197	36*
November	232	—	—	—	—	—	307	204	34
December	262	—	—	—	—	—	329	210	32
1970									
January	294	300	—	284	—	—	352	214	32
February	325	290	301	315	—	—	15	216	33
March	353	—	—	—	—	—	35	215	36
April	24	—	43	—	43	62	56	212*	39†
May	53	—	—	—	80	98	77	209	43
June	83	49	61	—	117	134	98	207	47
July	111	—	—	—	152	169	117†	207	50
August	141	156	168	176	186	201	137	210	52
September	171	—	—	164	216	227	157	215	52
October	200	—	—	—	234	—	176	221†	51
November	231	—	—	—	—	220	196	228	48*
December	262	271	282	—	223	233	215	234	46
1971									
January	293	—	270	278	246	261	234	240	44
February	324	293	—	—	279	295	253	244	45
March	352	—	358	317	311	329	270	246	48
April	23	32	—	—	347	5	289	246	52
May	52	23	26	37	23	41	306	243*	57†
June	81	—	—	—	61	79	319	239	60
July	110	—	132	146	—	—	324	237	63
August	140	157	160	—	—	—	316*	237	65
September	170	—	152	—	—	—	312	240	66
October	199	—	—	—	—	231	317	245	66
November	230	—	251	263	251	269	333	251†	64*
December	261	—	—	—	288	307	351	258	60

000 *Evening.* 000 *Morning.* † *Conjunction with Sun.* * *Opposition to Sun.*

TABLE 8 (continued)

Planet	Sun	Mercury			Venus		Mars	Jupiter	Saturn
Date	13th	3rd	13th	23rd	13th	28th	13th	13th	13th
1972									
January	292	259	272	—	326	344	12	265	59
February	324	—	—	—	3	21	32	271	59
March	353	—	12	—	37	54	51	276	61
April	24	—	4	6	70	83	71	278	64
May	53	16	31	—	91	95	91	278	67†
June	82	—	—	111	—	78	110	276*	72
July	111	126	137	142	76	83	129	271	75
August	141	—	—	132	95	109	149†	269	78
September	171	—	—	—	125	142	169	269	80
October	200	—	—	231	159	177	188	272	80
November	231	244	252	—	196	214	208	277	79
December	262	—	242	252	233	252	228	283†	76*
1973									
January	293	—	—	—	272	290	249	290	73
February	325	—	—	351	—	—	271	297	73
March	353	—	—	—	—	—	291	304	74
April	23	348	356	8	—	—	312	309	77
May	52	—	—	—	—	—	333	312	80
June	82	—	105	116	99	118	355	312	84†
July	111	122	—	—	136	154	14	310	88
August	140	—	123	—	174	192	32	305*	91
September	170	—	—	—	210	228	40	302	94
October	200	212	225	234	245	261	36*	302	95
November	231	—	—	222	278	292	25	305	94
December	261	232	—	—	304	312	26	310	92*
1974									
January	293	—	—	—	311	—	37	317	88
February	324	—	341	—	296	300	52	324†	87
March	352	—	327	335	308	320	68	331	88
April	23	347	3	—	336	353	86	337	89
May	53	—	—	—	9	26	104	343	93
June	82	96	102	—	44	62	122	347	96†
July	111	—	—	100	79	97	141	348	100
August	140	—	—	—	118	136	161	346	104
September	170	—	191	205	—	—	180	342*	106
October	200	215	220	—	—	—	200†	338	108
November	231	—	212	—	—	—	221	338	109
December	261	—	—	—	—	—	241	340	108

000 *Evening.* 000 *Morning.* † *Conjunction with Sun.* * *Opposition to Sun.*

TABLE 8 *(continued)*

Planet	Sun	Mercury			Venus		Mars	Jupiter	Saturn
Date	13th	3rd	13th	23rd	13th	28th	13th	13th	13th
1975									
January	293	—	—	321	309	327	*264*	345	104*
February	324	—	—	*312*	347	6	*287*	352	102
March	352	*316*	*326*	*339*	22	41	*308*	359†	102
April	23	—	—	—	60	77	*331*	6	103
May	52	—	75	81	95	111	*354*	*13*	106
June	81	—	—	—	128	142	*17*	*19*	109†
July	110	*80*	*91*	—	153	160	*38*	23	113
August	140	—	—	—	160	—	59	25	*117*
September	170	186	197	205	*145*	*147*	76	23	*120*
October	199	—	—	*192*	*155*	*168*	90	19*	*122*
November	230	—	—	—	*184*	*200*	92	15	*123*
December	261	—	—	—	*217*	*235*	*83**	14	*123*
1976									
January	292	301	310	—	*254*	*272*	74	16	*121**
February	324	*294*	*297*	*309*	*291*	*310*	78	21	118
March	353	*320*	—	—	*327*	*346*	88	27	116
April	24	—	—	53	*6*	*24*	103	34†	116
May	53	—	—	—	—	—	118	*42*	118
June	82	—	*59*	*71*	—	—	136	*49*	121
July	111	—	—	—	—	—	154	55	125†
August	141	—	165	177	157	175	173	59	*129*
September	170	186	—	—	195	213	193	*61*	*132*
October	200	—	*184*	—	231	249	213	*60*	*136*
November	231	—	—	—	269	287	234†	57*	*137*
December	262	—	—	291	305	322	*256*	52	*137*
1977									
January	293	—	—	*280*	340	356	*279*	50	*135*
February	325	*290*	*303*	—	10	20	*303*	53	*133**
March	353	—	—	—	25	22	*325*	56	131
April	23	—	*42*	—	—	7	*349*	62	130
May	52	—	—	*38*	*12*	*22*	*13*	68†	132
June	82	*50*	—	—	*36*	*51*	*36*	76	134
July	111	—	—	143	*67*	*83*	*57*	83	137
August	140	158	167	171	*102*	*119*	78	89	141†
September	170	—	—	*162*	*138*	*156*	97	94	*145*
October	200	—	—	—	*175*	*194*	*113*	96	*148*
November	231	—	—	260	*214*	—	126	95	*150*
December	261	273	—	—	—	—	132	92*	*151*

000 *Evening.* 000 *Morning.* † *Conjunction with Sun.* * *Opposition to Sun.*

Table 8 (continued)

Planet	Sun	Mercury			Venus		Mars	Jupiter	Saturn
Date	13th	3rd	13th	23rd	13th	28th	13th	13th	13th
1978									
January	293	*262*	*269*	*282*	—	—	*124**	88	*151*
February	324	—	—	—	—	—	113	86	*148**
March	352	—	—	21	—	24	112	86	146
April	23	—	—	—	43	62	121	90	144
May	52	*18*	*26*	*39*	80	98	134	95	144
June	82	–	—	—	118	135	150	102†	146
July	111	121	136	146	152	169	167	*109*	149
August	140	152	—	—	186	201	185	*116*	152†
September	170	*144*	—	—	215	227	206	*122*	156
October	200	—	—	—	232	232	226	*126*	160
November	231	241	253	260	—	218	248	*129*	163
December	261	—	—	250	222	232	270	*128*	164
1979									
January	293	*262*	—	—	*246*	*261*	295†	*125**	164
February	324	—	—	—	*279*	*296*	*318*	121	*162*
March	352	359	—	—	*311*	*329*	*341*	119	160*
April	23	—	*358*	*6*	*348*	*5*	*5*	120	158
May	52	*18*	—	—	*23*	*42*	*28*	122	158
June	82	—	—	*114* ➤	*62*	*80*	*50*	127	159
July	110	127	133	—	—	—	*72*	134	162
August	140	—	*123*	*132*	—	—	*93*	140|	165
September	170	—	—	—	—	—	*113*	*147*	168
October	199	—	220	233	—	231	*131*	*153*	172†
November	230	243	—	—	251	270	*146*	*157*	175
December	261	*231*	*240*	—	289	307	*158*	*159*	*176*
1980									
January	292	—	—	—	326	345	*165*	*160*	178
February	324	—	—	350	4	22	*160**	*156**	177
March	353	—	—	*338*	38	54	149	153	*175**
April	24	*346*	*358*	*13*	70	83	146	150	172
May	53	—	—	—	90	92	153	151	171
June	83	94	106	113	—	*76*	166	153	172
July	111	—	—	—	*75*	*82*	181	158	174
August	141	*112*	—	—	*95*	*109*	200	164†	176
September	171	—	—	201	*126*	*142*	220	*170*	180†
October	200	215	225	229	*159*	*177*	241	*177*	*184*
November	231	—	—	*223*	*196*	*213*	264	*183*	*187*
December	262	—	—	—	*233*	*252*	286	*187*	*189*

000 *Evening.* 000 *Morning.* † *Conjunction with Sun.* * *Opposition to Sun.*

The Moon's Age

The Moon's age is measured in days from new moon; it is 7 days at first quarter, 15 days at full moon, and 22 days at third quarter. Table 9 is based on a simple approximate formula given by P. Harvey in the *Journal of the British Astronomical Association* in July 1941, and lists the Moon's age at the zeroth day of each month from January 1961 to December 1990.

How to Find the Moon's Age

Method A. From Table 9 (good from 1961 to 1990).

Enter Table 6 with the year and month. To the table entry add the number of the day of the month to give the Moon's age on that date. (If the sum exceeds 29, subtract 30).

The error is usually about 1 day, occasionally 2 days.

Example: What is the Moon's age on Christmas Eve, December 24, 1976?

For December 1976, Table 9 gives the entry 9. Thus the Moon's age on December 24 is 9 + 24 − 30 = 3 days. The Moon will be new and visible in the evening sky.

Note: For observations to be made in the evening, especially in the Americas, add 1 day to the age of the Moon determined from the table, since the dates here are for midnight at Greenwich, England.

Method B. For dates not included in Table 9.

Harvey's formula is usually accurate to within a day (or occasionally two) over the Christian Era, using the present Gregorian calendar. The calculation is as follows:

	For Dec. 24 1976
Divide the year number by 19; keep only the remainder	(0)
Multiply the remainder by 11	0
Add ⅓ of the century number excluding fractions	+ 6
Add ¼ of the century number excluding fractions	+ 4
Add the number 8	+ 8
Subtract the number of the century	− 19
Add the number of the month, beginning with March = 1 (February = 12 and January = 11 of the previous year)	+10
Add the day of the month	+24
Sum	33
Subtract multiples of 30	− 30
Age of the Moon (days)	3

TABLE 9.   *The Moon's age on the zeroth day of each month (January 1961 to December 1990).*

	Jan	Feb	Mar	Apr	May	Jun	Jul	Aug	Sep	Oct	Nov	Dec
1961	13	14	14	15	16	17	18	19	20	21	22	23
1962	24	25	25	26	27	28	29	0	1	2	3	4
1963	5	6	6	7	8	9	10	11	12	13	14	15
1964	16	17	17	18	19	20	21	22	23	24	25	26
1965	27	28	28	29	0	1	2	3	4	5	6	7
1966	8	9	9	10	11	12	13	14	15	16	17	18
1967	19	20	20	21	22	23	24	25	26	27	28	29
1968	0	1	1	2	3	4	5	6	7	8	9	10
1969	11	12	12	13	14	15	16	17	18	19	20	21
1970	22	23	23	24	25	26	27	28	29	0	1	2
1971	3	4	4	5	6	7	8	9	10	11	12	13
1972	14	15	15	16	17	18	19	20	21	22	23	24
1973	25	26	26	27	28	29	0	1	2	3	4	5
1974	6	7	7	8	9	10	11	12	13	14	15	16
1975	17	18	18	19	20	21	22	23	24	25	26	27
1976	28	29	0	1	2	3	4	5	6	7	8	9
1977	10	11	11	12	13	14	15	16	17	18	19	20
1978	21	22	22	23	24	25	26	27	28	29	0	1
1979	2	3	3	4	5	6	7	8	9	10	11	12
1980	13	14	14	15	16	17	18	19	20	21	22	23
1981	24	25	25	26	27	28	29	0	1	2	3	4
1982	5	6	6	7	8	9	10	11	12	13	14	15
1983	16	17	17	18	19	20	21	22	23	24	25	26
1984	27	28	28	29	0	1	2	3	4	5	6	7
1985	8	9	9	10	11	12	13	14	15	16	17	18
1986	19	20	20	21	22	23	24	25	26	27	28	29
1987	0	1	1	2	3	4	5	6	7	8	9	10
1988	11	12	12	13	14	15	16	17	18	19	20	21
1989	22	23	23	24	25	26	27	28	29	0	1	2
1990	3	4	4	5	6	7	8	9	10	11	12	13

Suggested Reading

Abell, George. *Exploration of the Universe.* New York: Holt, Rinehart & Winston, 1964. (Textbook.)

Asimov, Isaac. *Kingdom of the Sun,* rev. ed. New York: Schuman, 1963. (Popular study of Solar System.)

Asimov, Isaac. *Universe.* New York: Walker, 1966. (Popular account of astronomy.)

Baker, Robert H. *Astronomy.* Princeton: Van Nostrand, 1964. (Textbook.)

Baldwin, Ralph B. *The Moon.* New York: McGraw-Hill, 1965. (Popular study of meteoritic impact on the Moon.)

Baldwin, Ralph B. *The Measure of the Moon.* Chicago: Univ. of Chicago Press, 1963. (Study of meteoritic impact on the Moon.)

Clarke, Arthur C. *Interplanetary Flight,* 2 ed. New York: Harper Bros., 1960. (Popular description of mechanics of space flight.)

Clarke, Arthur C. *The Exploration of Space.* New York: Harper Bros., 1951. (Popular account.)

Firsoff, V. A. *Strange World of the Moon.* New York: Basic Books, 1959. (Popular account.)

Gamow, George. *Planet Called Earth.* New York: Viking, 1963. (Popular study of Earth.)

Hawkins, Gerald S. *Meteors and Comets.* New York: McGraw-Hill, 1964. (Popular account.)

Hawkins, Gerald S., and White, J. B. *Stonehenge Decoded.* New York: Doubleday, 1965. (Popular account.)

Hawkins, Gerald S. *Splendor in the Sky.* New York: Harper & Row, 1961. (Popular account of astronomy.)

List compiled courtesy of E. Nelson Hayes.

Hess, W. M., Menzel, D. H., and O'Keefe, J. A. eds. *The Nature of the Lunar Surface.* Baltimore: Johns Hopkins Press, 1966. (Symposium on lunar space results.)

Hoyle, Fred. *Frontiers of Astronomy.* New York: Harper & Row, 1955. (Popular account.)

Hoyle, Fred. *Astronomy.* New York: Doubleday, 1962.

Hynek, Joseph A., and Anderson, Norman D. *Challenge of the Universe.* New York: McGraw-Hill, 1962.

Jackson, Joseph H. *Pictorial Guide to the Planets.* New York: Crowell, 1965.

Jones, H. Spencer. *Life on other Worlds,* rev. ed. New York: New American Library, 1956.

Koestler, Arthur. *The Sleepwalkers.* New York: Macmillan, 1959. (History of astronomy through Newton.)

Larousse. Encyclopedia of Astronomy, 2nd rev. ed. New York: Putnam, 1959. (Extensive account of astronomy.)

Leonard, Jonathan, and Sagan, Carl. *Planets.* Morristown, N.J.: Silver Burdett, 1966.

Ley, Willy. *Ranger to the Moon.* New York: New American Library, 1965.

Ley, Willy. *Watchers of the Sky.* New York: Viking, 1963.

Ley, Willy. *Mariner IV to Mars.* New York: New American Library, 1966.

Lundquist, Charles A. *The Physics and Astronomy of Space Science.* New York: McGraw-Hill, 1966.

Markov, Aleksandr V., ed. *The Moon: A Russian View.* Chicago: Univ. of Chicago Press, 1962. (Advanced summary.)

Menzel, Donald. *Our Sun,* rev. ed. Cambridge, Mass: Harvard Univ. Press, 1959. (Popular account.)

Menzel, Donald. *Field Guide to the Stars and Planets.* Boston: Houghton Mifflin, 1963. (For popular use.)

Moore, Patrick. *The Planet Venus.* London: Faber & Faber, 1959. (Pre-space account.)

Moore, Patrick. *Survey of the Moon.* New York: Norton, 1963.

Page, Thornton, and Page, Lou Williams. *Neighbors of the Earth.* New York: Macmillan, 1965.

Page, Thornton, and Page, Lou Williams. *Wanderers in the Sky.* New York: Macmillan, 1965.

Page, Thornton, and Page, Lou Williams. *Origin of the Solar System.* New York: Macmillan, 1966.

Payne-Gaposchkin, Cecilia. *Introduction to Astronomy.* New York: Prentice-Hall, 1954. (Textbook.)

Richardson, Robert S. *Exploring Mars.* New York: McGraw-Hill, 1954.

Shklovskii, I. S., and Sagan, Carl. *Intelligent Life in the Universe.* San Francisco: Holden-Day, 1966. (Extensive popular account.)

Struve, Otto, et al. *Elementary Astronomy.* New York: Oxford Univ. Press, 1959. (Textbook.)

Vaucouleurs, Gerard de. *Discovery of the Universe.* New York: Macmillan, 1957. (History of astronomy to 1956.)

Watson, Fletcher. *Between the Planets.* Cambridge, Mass: Harvard Univ. Press, 1956. (Comets and meteors.)

Index